AF559762

Milk Testing a Laboratory Control of Milk

Milk Testing a Laboratory Control of Milk

Dr. Pradeep Kumar Choda

Milk Testing a Laboratory Control of Milk

ISBN 978-93-5111-794-0

Published in 2016 in India by

RANDOM PUBLICATIONS

4376-A/4B, Gali Murari Lal, Ansari Road
New Delhi-110 002
Phone : +9111-43580356, 011-23289044, 011-43142548
e-mail: sales@randompublications.com,
info@randompublications.com, randomexports@gmail.com

Type Setting by : Friends Media, Delhi-110089
Digitally Printed at : Replika Press Pvt. Ltd.

Preface

Dairy foods and ingredients in general and fermented milk products in particular are well recognized for their nutritional and therapeutic attributes and as functional foods. The market of functional foods, in particular fermented milk products, is expected to triple in next five years world over because fermented milks possess many advantages like easy production, better keeping quality and nutritive value, easier digestibility, pleasing flavour and taste as well as strong therapeutic potentials.

Testing of raw milk is required to determine the composition (fat, protein), the bacterial and biological levels (Total Bacteria Count and Total Somatic Cell Count) as well as to determine the inhibitory substances such as dirt, antibiotics, disinfectants etc.

Milk is a fluid secreted by the mammary glands of healthy cows of the bovine species during the usual lactation period by means of complete and regular milking. It is a concentrated food, designed to initially protect the newly born animal by supplying concentrated antibodies to counter disease (passive immunity), until the immune system of the young animal is fully functional, and thereafter also to ensure rapid growth until the young are more independent.

In this book the author has endeavoured to include all the necessary information for the routine control of milk (raw, pasteurized and sterilized) and also (hopefully) of cream.

– Author

Contents

1

Quality Control of Pasteurised Milk

When milk is pasteurised at 63°C for 30 min in batch pasteuriser or 72°C for 15 seconds in heat exchanger, continuous flow pasteurisers, all pathogenic bacteria are destroyed, there by rendering milk safe for human consumption. Simultaneously various enzymes present in milk, and which might affect its flavour, are destroyed. In order to determine whether or not milk has been adequately pasteurised, one of the enzymes normally present in milk phosphatase, is measured. A negative phosphatase result indicates that the enzyme and any pathogenic bacteria have been destroyed during pasteursation. If it is positive, it means the pasteurisation process was inadequate and the milk may not be safe for human consumption and will have a short shelf life.

- Test tubes
- 5 mls pipettes
- 1 ml pipettes
- 100 ml volumetric flask
- 500 ml volumetric flask
- water bath at 37°C

Note: All glassware must be rinsed, cleaned, rinsed in chromic acid solution and boiled in water for 30 min. Buffer solution: Is mixed by 0.75g anhydrous sodium carbonate and 1.75g Sodium bicarbonate in 500 ml distilled water.Buffer-substrate solution: Place 0.15 gofdi-sodium paranitrophenylphosphate(the substrate)into a clean 100ml measuring cylinder. Add the buffer solution to make to 100 ml mark. Store this buffer-substrate solution in a refrigerator and protected against light.

It should not be used after one week. Prepare a fresh stock. Procedure: Pipette 5mls buffer-substrate solution into a test tube, stopper and warm the solution in the water bath at 37°C. Add to the test tube 1ml of the milk to be tested, stopper and mix well and place in water bath at 37°C. Prepare a blank sample from boiled milk of the same type as that undergoing the test.

Incubate both the test samples and the blank sample at 37°C for 2hrs. After incubation, remove the tubes and mix them thoroughly. Place one sample against the blank in a Lovibond comparator" all purposes" using A.P.T.W. disc and rotate the disc until the colour of the test sample is matched and read the disc number.

Interpretation:

Disc Reading after 2 hrs incubation at 37ºC	Remarks
0-10	Properly pasteurised
10-18	Slightly under pasteurised
18-42	Under Pasteurised
> 42	Not Pasteurised

DEFINITION OF QUALITY OF SKIMMED MILK POWDERS

Skimmed milk powder can be produced by different methods and is classified by the American Dry Milk Institute (ADMI) in different categories according to non-denatured soluble protein content as “low-heat”, “medium-heat”, and “high-heat” powder. Two physico-chemical criteria can be applied to classify powders on the basis of their denaturation by heat, i.e. solubility value and soluble protein.

SOLUBILITY VALUE

The method used (modified ADMI method) consists of dissolving in cold form a given quantity (5 g) of powder in 35 ml distilled water. After shaking followed by centrifugation for 10 minutes at 1 800 rpm, the supernatant is removed; the deposit is put back in suspension in 40 ml distilled water and is subjected to further centrifugation for 10 minutes. The insoluble portion is then oven-dried at 95°C for 15 hours. The percentage of insoluble material can be calculated from the weight of the dry residue and consequently the solubility value.

Insoluble Protein Value

The method here consists in measuring the nitrogen content of the non-soluble part of a reconstituted milk under standard conditions (2 g powder in 80 ml distilled water). After shaking and centrifugation for 15 minutes at 3 000 rpm, the supernatant is removed and the sediment put back in suspension in 80 ml water. After further centrifugation the residue of insoluble matter is mineralised and the nitrogen is titrated by the Kjeldahl method. The insoluble protein value is calculated by comparing the nitrogen content of the residue to the total nitrogen content of 2 g powder.

DEFINITIONS OF QUALITY OF ANHYDROUS MILK FAT (AMF)

Anhydrous milk fat can also show differences in quality due to the raw material, manufacturing process and length of storage. Two chemical values are commonly used to describe the quality of this fat, i.e. peroxide value and acid value.

Peroxide Value

Measurement of the peroxide value has been used to evaluate the degree of oxidation. The method consists in dissolving the fat in 30 ml of a chloroform-acetic acid mixture. A small quantity of potassium iodide is added. The peroxide releases the iodine which is then titrated by sodium thiosulfate in a starch paste. The results are expressed in milliequivalents (meq.) per 1 000 g fat.

Acid Value

The degree of lipolysis of fat is expressed by means of the acid value. The method consists in dissolving the fat in an ether-alcohol mixture, then neutralising the acidity of the fat by 0.1 N caustic soda in phenolphtalene. The results are expressed in ml of 0.1 N soda per 100 g fat.

THE INFLUENCE OF RAW MATERIAL QUALITY ON THE TASTE OF RECOMBINED MILK

A study was undertaken at the Polytechnic Institute of Lorraine (France) to determine accurately the effects on the taste of milk recombined from milk powder and butterfat of varying quality, and to establish the maximum and minimum limits of reincorporation of recombined milk in local raw milk. The raw materials used in this study came from batches which were freshly manufactured during 1980. Three skimmed milk powders, each with a different level of denaturation, were used.

Table. Chemical Characteristics of Quality of Powders Used.

Powder	Solubility (%)	Non-soluble protein (%)
Slightly denatured	99.340	0.457
Moderately denatured	99.016	0.622
Highly denatured	97.292	1.668

In order to have a range of AMF qualities it was necessary to modify the original fat by oxidation or by controlled lipolysis. The characteristics of this basic substrate are as follows: moisture content 0.2 percent (measured by Karl-Fisher method); peroxide value 40 mg per 1 000 g fat; acid value 1.22 ml per 100 g fat. A portion of this basic sustrate was subjected to oxidation while another was subjected to lipolysis under the following conditions.

Oxidation of AMF

The unmodified fat was dissolved at 40°C and kept at that temperature during the whole oxidation period. The latter process continued for about 12 hours under the action of ultraviolet rays from a lamp placed about 30 cm from the surface of the fat. Under these conditions the peroxide value rose to 7.5 meq. per 100 g fat. This highly oxidized fat was then used in a mixture adjusted with unmodified fat to obtain different levels of oxidation. It was immediately added to milk that had been reconstituted or stored at 20°C away from light.

Lipolysis of AMF

The unmodified and previously dissolved fat was enzyme-lipolized (0.5 – 8 g lipase per 100 g fat). After shaking, the fat was kept at 36°C for periods varying from 24 h to 7 days in order to obtain different degrees of lipolysis. The fat was then used immediately for recombining the milk. The raw milk (non-fat dry matter content 8.5 percent and fat content 3.5 percent) used for the purpose was subjected to pasteurisation at 78°C for 20 seconds. The recombined milk and the mixtures were made as follows: Skimmed milk powder was poured slowly and without shaking in distilled water, heated previously to 60°C, at a rate of 89.5 g powder to 876.5 g water. Shaking was continued for three minutes and the reconstituted milk was stored at + 5°C for 14 hours to stabilize the protein hydration level. In this way 966 g of reconstituted milk was obtained.

The milk was then heated to 60°C and the fat dissolved at the same temperature was incorporated at the rate of 34 g per 966 g reconstituted milk. The mixture was homogenized in a 2-stage homogenizer (APV Junior) at 60°C and at 185 kg/cm^2 pressure at the first stage and 75 kg/cm^2 at the second stage. A recombined milk was then obtained having 3.5 percent fat and 8.5 percent non-fat dry matter. The milk recombined in this way was then mixed in different proportions with the pasteurised fresh milk.

Given the equal dry matter and fat matter composition of these different products, the composition of the mixed milk did not have to be changed; the latter was submitted to the tasting panel within a maximum period of two hours. The tasters were persons qualified and trained to detect variations of fat taste in the recombined milk and of powder taste in the reconstituted milk. Their selection was made by a three-way test and by a classification test. The sensory analysis locations were equipped as specified by French standards. Between 5 and 10 samples were submitted per session and the average time of each testing was 20 minutes for each of the analysts. The methodology was that of assigning points on a scale increasing with the intensity of each of the three sought-after flavours.

Table. Points Scale.

Points to be assigned	Extent of defect
1	Nil
2	Uncertain
3	Perceptible
4	Considerable
5	Marked
6	Very marked

To apply these results it was arbitrarily agreed that the threshold for perception of an unpleasant taste for 50 percent of the tasters should be set at an average number of 3 points. Only one constituent at a time was changed during the test so as to assess the effect of the quality of the milk powder and the anhydrous fat on the taste of the reconstituted milk. The analysis of milk powder quality and together with the proportions used in the mixture to reach the level at which the defect called "powder taste" appears.

Table. Relationship between Chemical Quality of Powders and Permissible Quantity of Fresh/Recombined Milk Mixture (Average Value on Points Scale: 3).

Powder quality	Solubility (%)	Non-soluble protein (%)	Extent of mixture at level 3 (%)
Slightly denatured	99.340	0.457	59.0
Moderately denatured	99.016	0.622	53.5
Denatured	97.292	1.668	45.5

By setting 3 points as the threshold of unpleasant taste perception, it was found that with a high temperature powder mixing should not exceed 45 percent if adequate organoleptic quality is to be maintained. On the other hand, with powder prepared at a low temperature it seems that up to 59 percent of recombined milk can be added. Furthermore, the extent of the defect as a ratio of the percentage of recombined milk included in raw milk. According to the extent of powder denaturation, the panel described the taste defect of recombined milk not mixed with raw milk as marked or considerable.

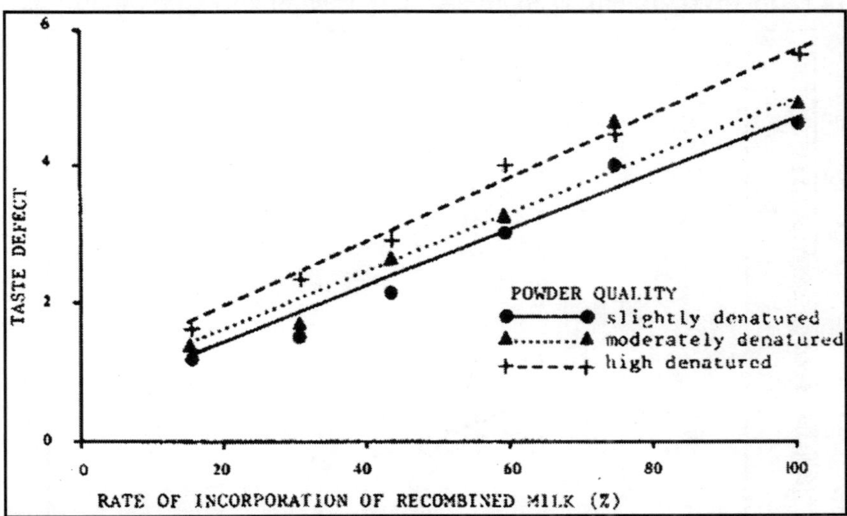

Fig. Influence of Rate of Incorporation of Recombined and Raw Milk on Defect Referred to as "Powder Taste".

The influence of anhydrous milk fat quality is demonstrated by the appearance of an oxidized taste and a hydrolytic rancid taste. The extent of the oxidized taste shows a linear correlation with the amount of recombined milk added. As the peroxide value rises, the defect increases.

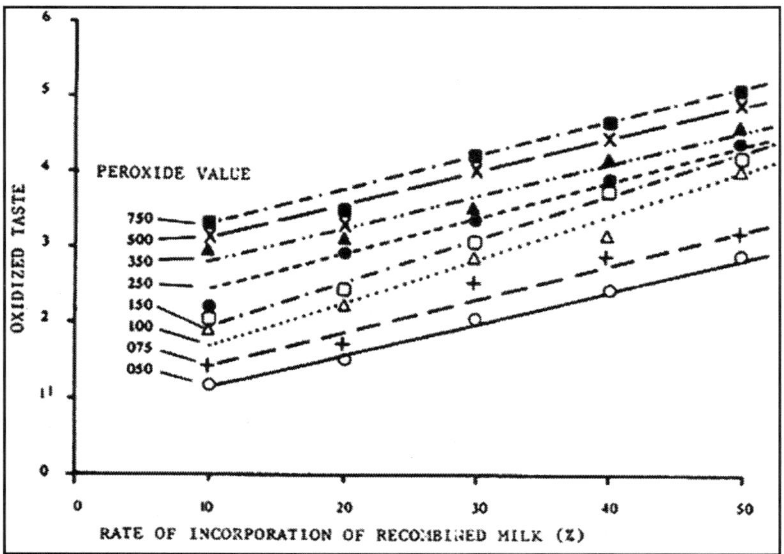

Fig. Influence of Fat Oxidation and of Rate of Mixture of Recombined and Raw Milk on the Defect Called "Oxidized Taste".

There is a correlation between the fat peroxide value and the acceptable level of re-incorporation of recombined with raw milk.

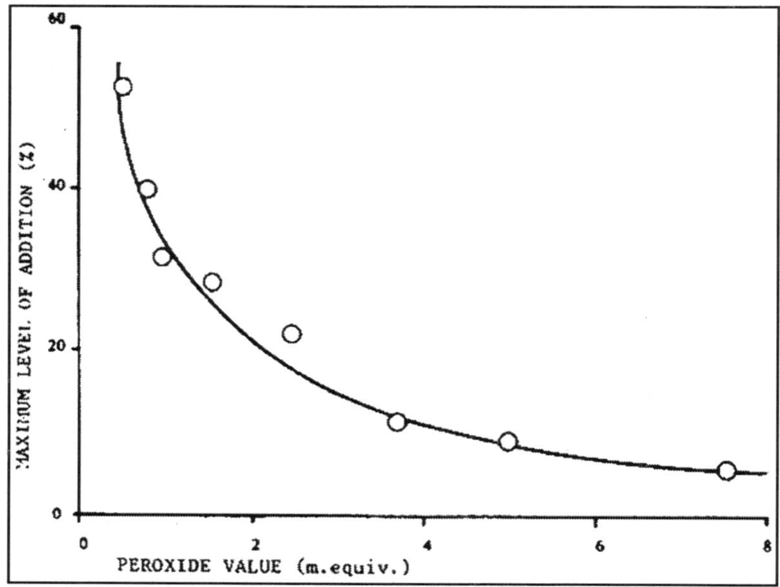

Fig. Maximum Level of Addition of Recombined Milk in Raw Milk (Level 3 of Sensory Analysis) as a Function of the Fat Peroxide Value.

In the case of the original (unmodified) fat, it seems therefore that more than 50 percent recombined milk can be added. On the other hand, in the case of a fat with a very high peroxide value (7.5 meq) the maximum possible level of addition is very low (less than 10 percent). These trials indicate that the measurement of peroxide value is useful in ascertaining the extent of the oxidized taste in milk. Statistical analysis of the findings has shown that the correlation is significant (r > 0.88).

Perception of a hydrolytic rancid taste depends on the incorporation rate of recombined milk with raw milk in a similar manner as perception of oxidized taste. In testing for the maximum and minimum degree of addition resulting in a detectable objectionable taste, a correlation may be seen between this rate and the acid value expressed in ml 0.1 N (caustic) soda. The correlation is linear within the limits of the experiment, in other words between acid values of 1.2 and 8 ml.

The correlation coefficient is highly significant (r = 0.99) and warrants the use of acid value measurement to predict the degree of the rancidity defect. For example, with fresh fat (acid value 1.2) it seems that up to 70 percent of recombined milk can be introduced in raw milk without changing its organoleptic character. On the other hand, if the fat is very rancid (acid value 8) the maximum level is very low (less than 10 percent).

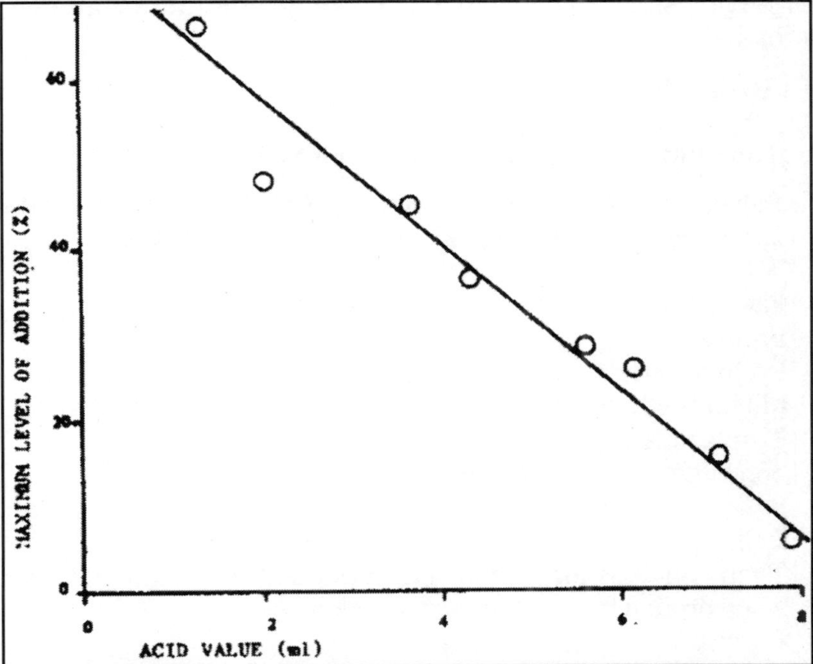

Fig. Maximum Level of Addition of Recombined Milk in Fresh Milk (Level 3 of Sensory Analysis) According to Fat Acid Value.

With the sensory method employed it was possible to identify the appearance levels of defective taste resulting from the use of raw materials with different degrees of spoilage. Thus use of a slightly denatured powder of the low temperature type warrants a higher rate of incorporation than with highly denatured powders. However, the differences remain small (45 to 59 percent under test conditions). It is pointless, therefore, even with a high quality powder, to try to produce a milk having a flavour equal to that of fresh milk.

Taste defects resulting from an alteration of fat have a much more marked influence on levels of mixing with fresh milk. It is necessary therefore to have complete control over the chemical quality of anydrous milk fat (its peroxide and acid values should be as low as possible). The generally accepted quality standards make it difficult to have recombined milk reincorporation percentage in excess of 40–50 percent.

It should also be pointed out that the group of tasters is not necessarily representative of the consumers from whom the recombined milk is intended. Because of food habits, differences in perception levels are likely so that the results given in this report should be interpreted and transposed with caution and the particular conditions of the respective population groups borne in mind. It would be useful to correlate these results with estimates based on surveys conducted among local populations.

PASTEURIZED VS. RAW MILK

Chemical and Physical Properties of Full and Skimmed Milk Powder

The important quality parameters for milk powder are microbiological quality, sensory characteristics and the following physicochemical properties:

- Water content
- Fat content
- Free fat
- Protein content
- Minerals content
- Titratable acid
- Solubility, reconstitutability
- Pouring behaviour
- Bulk density
- Thermal stress for skimmed milk powder (fraction of non-denatured whey proteins)
- Burnt particles
- Particle size distribution
- Residual oxygen level in the packaging

Table. Summary of the Physicochemical Properties of Milk Powder.

	Spray-dried powder	Drum-dried powder
Particle structure	Spherical particles, air inclusions	Compact, irregular shape, no air inclusions
Particle surface	Smooth, partly conduplicate	
Particle size	10–250 μm	
Bulk density [g/cm³]	0.50–0.70	0.3–0.5
Solubility, denaturation	Low denaturation of proteins ➔ good solubility	Higher degree of denaturation of proteins ➔ poor solubility
Heavy metal content requirements	Copper < 1.5 mg/kg Iron < 10.0 mg/kg	Analogous
Residual oxygen content in powders containing fat	≤ 0.01 ml O_2/g	
Browning caused by the Maillard reaction	Less	Stronger

The drying process has a central influence on the product characteristics. A distinction must therefore be made between spray-dried and drum-dried powder.

Milk powder should pour easily, like sand, without forming lumps. Pourability is adversely affected by incompletely crystallised lactose, which absorbs water. Pouring behaviour is also affected by particle size and shape, bulk density and electric charge.

In whole milk powder or powder from partially skimmed milk the milk fat may be changed by oxidation. This oxidation is aided by the presence of high oxygen concentrations in packaging, light and metal ions such as copper and iron and, to a lesser extent, other metals. The requirements for maximum copper and iron content. The exclusion of oxygen using vacuum packing or packaging with nitrogen in a protective atmosphere is important for the prevention of oxidation when packaging milk powders containing fat. Good management of preliminary heating during powder production reduces oxidation changes in the end product.

Burnt particles are produced by a strong Maillard reaction (reaction of sugars with amino acids – the building blocks of proteins) during the drying process if particles stay in the process for too long.

A certain amount of browning from the Maillard reaction is normal for milk powder. The Maillard reaction is stronger with drum drying than with spray drying. The Maillard reaction continues to operate during storage, so the storage conditions (temperature, period) are important for maintaining the quality of milk powder.

Production/ Technology

The basic steps in milk powder production are evaporation and drying. The production sequence.

Milk of high quality is selected. As the concentration of dissolved and dispersed substances in the milk increases during powder production, the milk must be of high stability or the proteins can become destabilised and break down. An acidity of < 7.5°SH should be required. There must be no contamination by metal ions, particularly copper. The bacterial count requirements should be met. Dirt particles are separated from the milk in the centrifugal cleaner, it is then cooled and kept refrigerated. Depending on the type of powder, some or all the milk fat is centrifuged off as cream.

The subsequent heat treatment has various objectives:

- To inactivate all the pathogenic bacteria and reduce the overall bacterial count
- To inactivate enzymes, particularly lipase
- To activate SH groups in ß-Lactoglobulin so as to increase the powder's oxidation stability during storage.

Fig. Abbildung 2: Ansicht Einer Eindampfanlage Für Milch.

"High-short" methods are preferred, as these achieve the desired effects more gently and form more antioxidative substances. Powder solubility is also improved when a high-short method is used. Heating often takes place for 15-30 seconds at 88-95°C, sometimes at temperatures up to 130°C.

The milk then passes to the evaporation plant. Water is evaporated gently under vacuum at temperatures of 45-75°C until a concentration of 33-35% or 40-50% of milk dry matter is obtained. The first range applies to subsequent drum drying, the second to subsequent spray drying.

In the evaporation plant multiple use ensures energy efficiency. Thanks to thin-film technology the milk and resultant milk concentrate stay only very briefly in the system at elevated temperature.

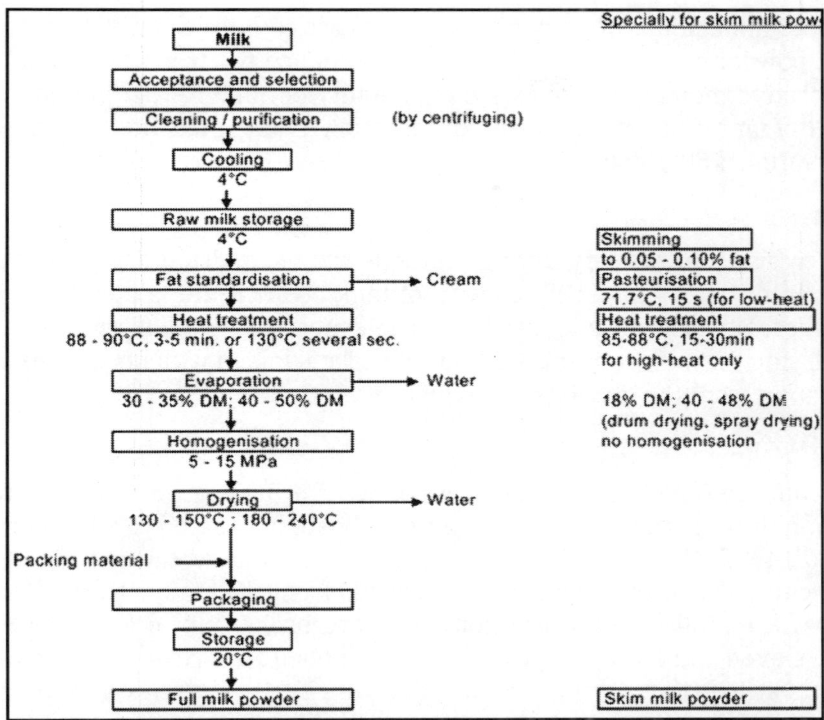

Fig. Abbildung 3: Fliess-Schema der Herstellung von Milchpulver.

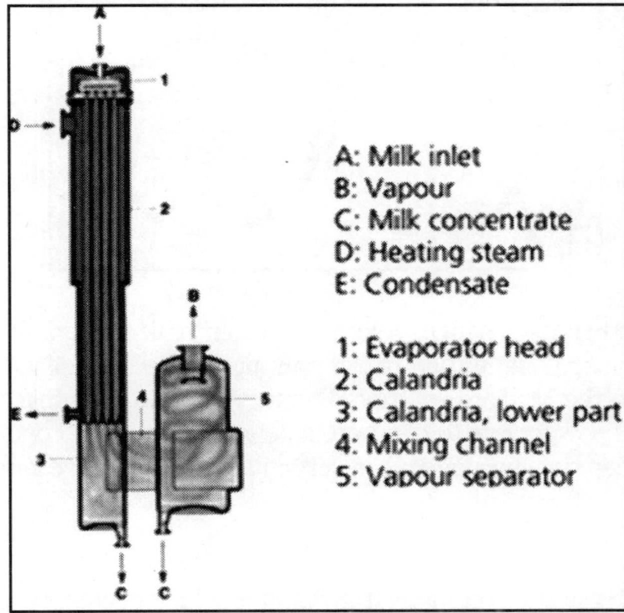

Fig. Diagram of a Falling Film Evaporator.

A certain amount of homogenisation is carried out during the production of milk powder containing fat. The aim is to reduce the free fat. Free fat with no protective membrane reduces the solubility of milk powder and increases the risk of fat oxidation. Following evaporation, homogenisation is carried out at pressures of 50-150 bar.

Drying

Two different primary drying methods are used, drum drying and spray drying. In drum drying a thin coating of milk concentrate is applied to a hot surface, causing a large amount of water to evaporate within a few seconds. In spray drying the milk concentrate is atomised and mixed with hot air, the water evaporating from the droplets.

Drum Drying

A thin coating of milk concentrate of 30-35% dry matter is applied to the revolving drying drums. These are internally steam-heated to a temperature of up to 145°C. Within less than 3 seconds a residual water content of only 4% is achieved and the dried milk is scraped from the drums by blades. The powder flakes fall into a worm conveyor, are broken up in a hammer mill, cooled, sieved and then packed. Drum drying plants are relatively compact in comparison to spray drying plants and give a better return on investment. The drum drying process produces flat, flake-like particles.

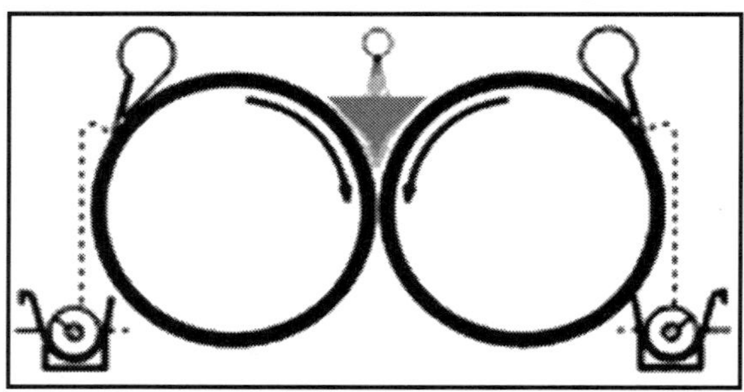

Fig. Diagram of a Drum Dryer.

Certain irreversible changes take place when milk is drum dried. The high temperatures cause denaturation of the milk proteins, caramelisation of lactose, browning due to the Maillard reaction. Drum-dried powder is less water-soluble than spray-dried powder due to the protein denaturation. The properties of drum-dried milk powder are advantageous in certain applications, for example chocolate production.

Spray drying

Spray drying is the commonest method used for drying milk powder.

Milk concentrate is sprayed into the top of a drying tower several storeys high in tiny droplets of 50-80 μm in diameter, using a spray nozzle atomizer or rotary atomizer.

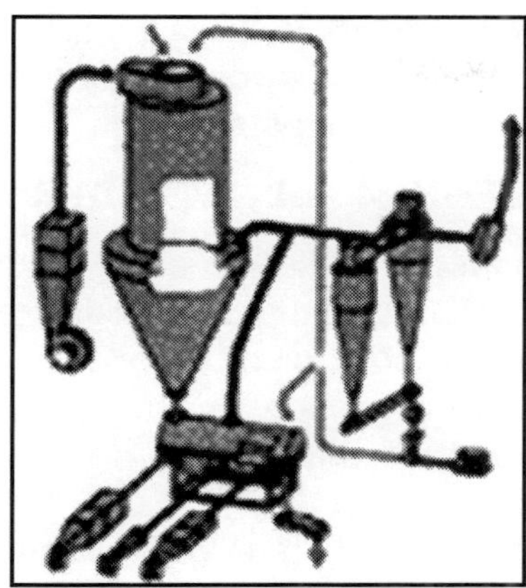

Fig. Two-stage Spray Dryer with a Further Fluid Bed Drying Stage Under the Tower.

Generally filtered hot air at a temperature of 150-300°C is also blown into the top of the tower. This facilitates gentle drying. The round droplets formed during atomisation retain their shape during the drying process, so the particles of spray-dried powder are spherical. Air trapped in the particles produces a lower bulk density. The water in the minute droplets evaporates rapidly, within 100th to 10th of a second. This causes rapid cooling of the particles and air. The maximum particle temperature on completion of drying is only 65-75°C. After discharge from the drying tower and any final drying in the fluid bed dryer, the powder is separated from the drying air in several steps, also with the aid of cyclones, and cooled with cold air.

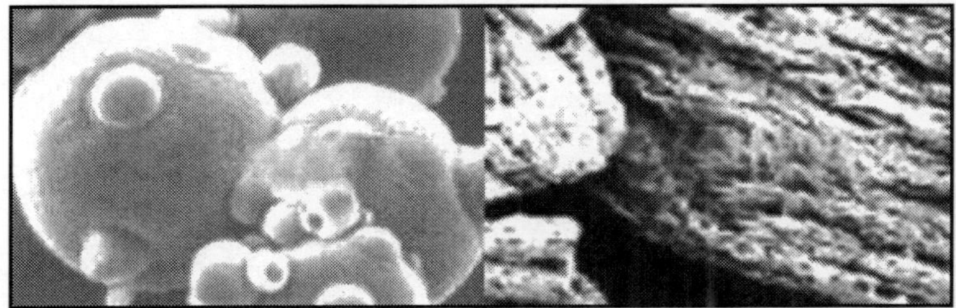

Fig. Microstructure of Milk Powder (Electron Micrograph), on the Left Drum-Dried, on the Right Spray-Dried; Particle Size: Drum-Dried Powder, Primary Particles Approx. 250–500 ym, Spray-Dried Powder, Primary Particles 25–50 ym Diameter. (Source: Livstek).

Packaging and Storage

Appropriate packaging is important in maintaining the quality of milk powder. The packaging must protect the milk powder from moisture, air, light and contamination, and of course logistical requirements must also be met. Use is often made of paper with a bitumen layer, multi-ply cartons or boxes with polyethylene liners, metal drums with polyethylene liners or cans with aluminium foil lids. Shelf life can be extended by the removal of oxygen using a protective gas atmosphere or vacuum packing. Milk powder is stored at ambient temperature. As whole milk powder is prone to fat oxidation, its storage life is shorter than that of skim milk powder.

SPECIAL FEATURES OF SKIM MILK POWDER

The flow sheet the differences between the production of skimmed milk powder and whole milk powder. All the cream is skimmed off, the heat treatment is reduced to gentle pasteurisation (low-heat) or extended to 15-30 min. (high-heat). The post-evaporation concentration of drum-dried skim milk is considerably lower than that of whole milk. However skim milk powder is generally produced by the spray method.

Low heat – high heat: Skim milk powder is classified by its heat stress and resultant protein denaturation. The production of low-heat powder is effected with the minimum possible heat stress to the milk, the concentrate and the powder.

Cold milk separators are used for low-heat powder, even for skimming. The dwell times at elevated temperature are kept as short as possible. Only gentle pasteurisation is used, and no intensive heat treatment. High-heat powder undergoes intensive heating, which denatures the maximum possible number of whey proteins. This is desirable for use in baked goods, for example.

Rapid cooling after drying is particularly important for skim milk powder because of its sensitivity to moisture and heat-associated reactions. Skim milk powder can be kept in hermetically sealed packs for over 1 year at 21°C. As it only contains very little fat, the oxidation risk is low.

Table. Classification of Skim Milk Powder by Heat Stress.

Quality	Pre-treatment of the milk	Non-denatured whey protein content WPNI*
Low heat	Cold milk skimming Pasteurisation 71.7°C, 15 s	>= 6 mg N/g
Medium heat		> 1.5 and < 6 mg N/g
High heat	85–88°C, 15–30 min	<= 1.5 mg N/g

Note:

WPNI = Whey protein nitrogen index

Instantising

The powder particles undergo various processes when milk powder is dissolved in water for reconstitution:

- Water absorption on the surface (wettability)
- Penetration of the water film at the particle surface (penetrability)
- Sinking in water (sinkability)

- Particle distribution without forming lumps (dispersibility)
- Dissolving of particles (solubility)

The object of instantising milk powder is to improve the speed and completeness of powder reconstitution. The methods below are successful in improving some of the above sub-processes at given powder solubility.

Fig. Abbildung 8: Mikrostruktur Von Agglomeriertem Milchpulver.

Agglomeration

Agglomeration causes the formation of voids between powder particles, allowing the easy and rapid penetration of water into these voids during reconstitution. This prevents a viscous layer forming around clusters of compact powder particles, as occurs in non-instantised powders. In non-instantised powders such a viscous layer prevents continued water penetration, thus slowing down reconstitution. The agglomeration process involves wetting the particle surface with steam, water or a mixture of both, the actual agglomeration, final drying, cooling and sifting to remove excessively fine and coarse particles.

There are basically two methods of effecting agglomeration:

- Primary agglomeration is carried out during spray drying by feeding fine powder back into the atomised mist of milk concentrate.
- Secondary agglomeration is carried out by re-wetting powder which is already dry.

Lecithinisation

This type of agglomeration is not enough for powders containing fat, as free fat on the surface of the powder prevents wetting. These powders are therefore sprayed with a lecithin solution in addition to agglomeration.

NEW FINDINGS - VARIOUS

The production of special-purpose milk powders is increasingly tailored to specific applications. Here use is only made of that part of the milk needed for a particular application. Thus there are highly protein-enriched milk protein powders or even protein-enriched milk powders with increased casein content or increased whey protein content. Evaporation and drying techniques are being optimised to improve product quality on the one hand and reduce energy consumption and overall drying costs on the other. The initial concentration step in skim milk production can be carried out by reverse osmosis, allowing concentration to approx. 25% dry matter. Further concentration is effected by evaporation. Reverse osmosis cuts energy costs.

Milk powder plants have to produce large volumes of powder in order to be financially viable, so various Swiss plants were closed down in the course of structural reorganisation. In 2001 there were still 6 in operation (Emmi Dagmersellen, Néstle Konolfingen, Nutritec Hochdorf, Swiss Dairy Food Sulgen, Thun and Lucens). Plants are becoming increasingly specialised and producing ever more special-purpose milk powders, e.g. tailor-made milk protein powder.

Use

Milk powder is a valuable ingredient in terms of both nutrition and technological function, and as such is in demand for a wide range of foodstuffs.

Whole Milk Powder

Whole milk powder dissolved in water is used as reconstituted milk. There is a large market for this, particularly in countries where little milk is produced. Considerable volumes of whole milk powder are blended with cocoa ingredients and sugar to produce quality milk chocolate. Further applications are other confectionery, biscuits, baked goods, coatings and various dairy products such as ice cream and processed cheese.

Skim Milk Powder

Skim milk powder has numerous applications. It is sold to the consumer direct in the form of reconstituted skimmed milk. Food manufacturers use it in dairy dessert products, ice cream, yoghurt, meat products, vegetarian alternatives to meat, for coatings, sauces, mayonnaise, instant breakfast drinks, etc. Powders spray dried especially gently from milk matched to breast milk prior to drying are used as the basis for infant foods/ baby milk.

MILK QUALITY

A critical prerequisite for the manufacture of quality dairy products is to start with milk of the highest quality. The milk quality criteria requested by milk processors and by customers generally are becoming more strict and rigorous. It is important that requests to milk producers for improved quality milk be accompanied by the necessary information required to produce such milk, particularly in light of new pressures from expansion of herds and the reality of reduced labour supply.

Further information on 'Milk Quality' is available at the links below. These include a 'Milking Equipment Cleaning Guide', which provides crucial information on milking machine washing procedures, chemical analysis of cleaning products and recommendations for the effective use of such products. There is also a link to a tool to assist with milk quality standards on-farm which has been made available by Teagasc.

This tool comprises a series of video clips showing critical stages of milking management (through to bulk tank milk storage) on both smaller and larger farms and on farms with adequate labour and those where technology can contribute. There is a further series of video clips produced by AHI which focuses on specific procedures pertaining to milk quality, e.g. antibiotic treatment of cows, drying-off procedures, liner changing, milking machine checks. There is also some information on suitable teat disinfection products and and flukicides that may be used.

Organic agriculture as an independent sector in Latvia exists from the 20th century nineties. The aim of organic agriculture is to create an integrated, human and environmentally friendly, economically well - balanced agricultural system, which is based on renewable raw materials of a local origin. Organic agriculture protects the cultivated plants from pests and illnesses, and provides agricultural animals with high quality feed. By developing of organic agriculture it is possible to reduce the negative influence of agricultural technology on environment and to improve the quality of obtained products, because the use of pesticides, organic compounds, exciters for growing, veterinary drugs and antibiotics are restricted in organic agriculture.

During the decrease of public trust in genetically modified products, as well as due to animal diseases, the demands for organic food and the interest in them increase. During the last years, the demand for organic food and the number of consumers, who assign more attention for high quality food and would like to know how it is produced, are significantly increasing. Organic agriculture is characterised by clear basic principles and the transparency of product origin, production and processing. There are all the necessary conditions in Latvia for production of qualitative livestock products for the internal market and as well as for export: land suitable for agriculture, multi-breed animal herds and ecological situation.

Scientists from different countries have very contradictory opinions about the chemical composition and quality of organic milk. The evaluation of chemical composition and quality of organic milk and its comparison with conventional milk has not been performed in Latvia by now. There is also lack of objective and grounded information about organic milk quality. Therefore it is necessary to evaluate the organic milk quality in order to provide consumers with safe nutrients and to help organic farmers to develop organic food production. After summarising theoretical judgements and experimental data from literature, the aim of the research work was set as follows - to determine and to evaluate the quality of organic milk.

The tasks of the research:

1. To investigate the chemical composition of organic milk;
2. To determine content of IgA, IgG, IgM, lactoferrin and lysozyme in organic milk;
3. To investigate somatic cell count and the microbiological quality of organic milk;
4. To determine the level of aflatoxin M_1 and heavy metals and trace elements (Cd, Zn, Pb, Cu, Fe) in milk;
5. To compare and to investigate the chemical composition and chemical contamination of organic milk and conventional milk.

Novelty of the research - the quality of organic milk in Latvia has been evaluated in complex, the concentration of antibodies in organic milk was determined for the first time. The potential differences of chemical composition of organic milk have been analysed and studied, taking as a basis feeding, namely, the data of composition of forage and cows' blood tests. Differences of chemical composition and quality of milk obtained from two various agricultural systems in Latvia have been explained.

The scientific importance of the research - the chemical composition, including the content of calcium, IgA, IgG, IgM, lactoferrin and lysozyme and urea was investigated, the microbiological quality of organic milk and somatic cell count was established, the evaluation of chemical contamination of organic milk was carried, out. The research results show that it is possible to get milk with normal chemical composition without using food additives. The economic significance of the research - the quality of organic milk is evaluated and this evaluation enables to make grounded conclusions about the significance of organic milk in consumers' diet.

APPROBATION OF THE RESEARCH WORK

The results have been presented at six international scientific conferences, symposiums and congresses in Latvia (Latvia University of Agriculture), Slovenia, Poland, Turkey, Hungary and Greece. The research results are reflected in six established and reviewed scientific publications in Latvian and English, three of them are published in the editions approved by international external and Latvian Council of Science. The list of publications is presented on page 6.

MATERIALS AND METHODS

The research work was being performed from October 2003 till November

2006. During 2006, the data were summarised and statistically processed, and the Doctor Thesis was prepared.

The researches were made at:

- Latvia University of Agriculture, Faculty of Food Technology, Scientific - Research Laboratory,
- Latvia University of Agriculture, Department of Chemistry, Laboratory of Water Analysis,
- Latvia University of Agriculture, Research Laboratory of Agronomy Analysis;
- Latvia University of Agriculture, Institute of Biotechnology and Veterinary Medicine „Sigra", Microbiological Laboratory of the Department of Veterinary Medicine;
- University of Latvia - Institute of Biology, The Laboratory of Biochemistry and Physiology of Animals,
- Riga Center of Reproduction,
- JSC „Sigulda station of breeding and artificial insemination",
- The Laboratories of National Diagnostic Centre at the Food and Veterinary Service.

The organic milk, conventional milk and milk obtained during transitional period (in further - the transitional period milk) were obtained from the farms "Lejasrembeni", "Jaunbiteni", „Kalna Gaurini", "Alejas", „Cemuri", which are located in Íeipenes rural district, Ogres region. Individual milk samples were collected from Riga, Cesis, Jelgava and Bauska region farmers.

Milk samples from different breeds of cows were used for the research: 61% of Latvian Brown, 2% of Holsteins Black and 37% crosses of Latvian Brown and Holsteins Black. There are sufficient researches about the influence of cow's breed on the milk composition and quality; therefore the factor of breed was not taken into consideration in this work. Considering the huge amount of the analysed milk samples, all investigated parameters were not detected for all milk samples. To evaluate the significant difference, the parameters were randomly arranged; parameters were detected for three duplications, the mean value of parameters was calculated.

The number of analysed milk samples and standards of analysis. The mean of protein, fat, lactose and somatic cell count in organic milk was determined and compared with the data for the same time period from Latvia's State Agency "Agricultural data centre". There is only one company, which processes organic milk in Latvia, and the company is located at K.eipenes rural district, Ogres region. The organic milk for this company is delivered by the following farms: "Lejasrembeni",

"Jaunbiteni", „Kalna Gaurini", "Alejas". Taking into consideration that the amount of organic milk obtained from the mentioned farms is 100% from all organic milk processed in Latvia, the analysed quality factors will enable to consider the organic milk quality in Latvia (in this connection the number of the analysed milk samples could be assumed as the general multitude). The data were processed by using the SPSS software package SPSS 11.0. and MS EXCEL.

RESULTS AND DISCUSSION

The Evaluation of Chemical Composition of Organic Milk

According to the data in literature, when changing the agricultural system (from conventional to organic), animal keeping conditions and feed composition, synthesis of the main chemical components in organic milk and their content in the product are influenced to a great extent.

The content of lactose, protein and fat in organic milk and conventional milk. The mean content of lactose in organic milk samples was 4.85±0.04%, it significantly differs ($p<0.05$) from those of conventional milk. The results of research relate to Olivo statement that the content of lactose in organic milk is significantly higher. At the same time, the research results do not conform to Byström's, Toledo - Alonzo's research results, where the authors affirm, that the content of lactose in organic and conventional milk have no significant difference. The content of lactose in milk is the least subjected to changes. The higher content of lactose in organic milk can be explained by the higher concentrations of sugar in feed grasses of organic farms. The mean content of protein in organic milk was 3.30±0.04%, which is not significantly different from conventional milk ($p>0.05$).

The research results relate with Haggar, Kristensen, Bystrom, Mogensen, Toledo-Alonzo and Ellis, that the content of protein in organic and conventional milk samples have no significant difference. The obtained results conform to the results of research where the total amount of proteins in blood of the cows treated by organic farming was less in comparison to those of conventional - 7.74±0.07% and 8.54±0.09% accordingly. The research results contradict with Olivo statement that the content of protein is higher in conventional milk samples. It is not strange, that different results have been reported regarding to the protein content of organic milk, since, organic milk production varies in feeding regimes. The reason for lower content of protein can be lack of sugar-rich juicy feed, which stimulates production of butyric acid used for protein synthesis. At the same time, according to Bluzmanis's research results, the sugar content in feed grasses obtained from organic farms is higher.

The mean content of fat was 4.98±0.08%; which is significantly higher than in conventional milk samples. The research results contradict with Kristensen, Mogensen, Toledo-Alonzo, Olivo statement that fat content is higher in conventional milk. Haggar, Bystrom, Ellis found no significant difference between the organic and conventional milk samples. The higher content of fat in organic milk could be explained with the differences in keeping and feeding conditions: high quality feed, well balanced and rich in cellulose, not chopped, in sufficient amount was available in organic farms; cows were always milked.

The content of urea is influenced by following factors: feed composition and dose, the weight of cows, milk yield, season, month and lactation period. The content of urea in organic milk samples. The urea content in organic milk samples ranged between 64.90 and 252.56 mg kg^{-1}. The mean content of urea in the analysed organic milk was 167.43±9.64 mg kg $^{-1}$, which fitted in the common limits set for milk - from 150 to 300 mg kg^{-1}. In 32% of organic milk samples the content of urea was for 29.6 mg kg–^{1} lower than the minimum limit - 150 mg kg^{-1}. Only 8% of analysed organic milk samples did not fit within Rajala - Schultz data limits from 103 mg kg^{-1} to 154 mg kg^{-1}. The research results relate with Toledo - Alonzo results, where the author determines a significant difference in the content of urea in milk from different agricultural systems. The difference between the urea content in organic and conventional milk samples can be explained by stricter rules regarding to the amount of mixed feed allowed in organic farms in comparison with the conventional farms - accordingly 40% and up to 65% of the dry matter daily. The higher amount of mixed feed allowed has a negative influence on paunch function and milk production.

The lower production intensity in organic farms, what Jemeljanovs' research results affirm, should be a possible explanation of the low milk urea level in organic milk found in this study. As livestock specialists suppose, the content of urea in milk should be till 150 mg kg^{-1}, otherwise it can burden insemination and reproduction. It means that from this point of view, the urea content in organic milk samples is normal.

The lower content of urea in 32% of organic milk samples could be explained by a low content of soluble nitrogen in feed and by the influence of seasonal changes in milk. The lowest content of urea, according to Godden 's data, is from April to June. The Dursts 's research shows that the normal content of protein that was found in organic milk - 3.18%, and low content of urea could be explained by the decreased content of total protein in feed. At the same time, according to Bluzmanis's research results, the total protein content in feed grasses obtained from organic farms is higher - 3.26±0.21% in comparison with the conventional - 2.99±0.30%.

Hojman notifies that the content of urea in milk correlates with fat content in milk. To verify this Hojman's statement, the regression analysis was used. The correlation coefficient was 0.246, which pointed to a weak connection. The factor - fat content in milk is not significant ($p>0.05$) for the changes of urea content in milk, it means that the model is not statistically significant. Within the framework of this research, the correlation between the urea and fat content was not found. The content of calcium in milk is influenced by feed components, season and the physiological factors of animal. The content of calcium in milk samples. The content of calcium in organic milk samples ranged between 20 and 25 mmol l^{-1}, the mean content of calcium was 21.90 ± 0.22 mmol l^{-1}. However, a statistically significant difference in the content of calcium between the organic and the conventional milk samples was not found - 20.80 ± 0.32 mmol l^{-1}. The mean content of calcium in milk samples from different agricultural systems was significantly lower if compared with the data from literature ($p<0.05$)-30 mol l^{-1}.

Gorbatova mentions that the content of calcium in milk obtained in summer period, is lower in comparison with milk samples obtained in winter, it could explain the decreased calcium content in milk samples taken in summer and autumn months, and it influences also the mean. A lower content of calcium could be caused by mastitis, the content of calcium can decrease then for 9%, The content of calcium in milk increases in the following order: from conventional to organic agriculture. No significant difference in content of calcium ($p>0.05$) in organic and conventional milk samples was found. The cows kept in organic agriculture did not get mineral additives, so boosting of their immunity was connected only with facilitation of natural self-regulation processes. By means of that it is possible to achieve the same results as in the conventional agriculture, where animals are treated, without prophylaxis of any disease.

Bluzmanis notes that the relation between calcium and phosphorus in blood serum from organic cows is better (2:1) then from the conventional ones (2.27:1). Within the framework of research the content of thiamin and riboflavin in milk was determined. The content of thiamin ranged between 0.20 to 0.32 mg l^{-1}. The content of thiamin in organic milk samples was for 34.1% lower if compared with the conventional milk and it was significantly lower ($p<0.05$) in comparison with the data from literature and conventional milk. The content of thiamin in conventional milk was 0.41 ± 0.01 mg l^{-1}, it conforms to the data from literature.

The parameters of content of riboflavin in milk. The content of riboflavin in organic milk ranged from 1.28 to 2.96 mg l^{-1}. The mean content of riboflavin in organic milk still was 1.70 ± 0.10 mg l^{-1} and it was for 35.8% lower than in conventional milk, so a statistically significant difference between the organic and conventional milk samples ($p<0.05$) was found. When comparing the parameters available in literature, no significant difference was established.The content of riboflavin in conventional milk was 2.65 ± 0.10 mg l^{-1}, which significantly differs from the data in literature ($p<0.05$). Many authors have pointed that the content of thiamin and riboflavin do not vary in different seasons and that the feed composition has no significant influence on it. However there is still a possibility that feed can influence the content of thiamin and riboflavin in milk, especially if the feed is rich in grain and flour. The decreased dosage of mixed feed, used in organic farms, can negatively influence the content of thiamin and riboflavin in organic milk. While decreasing the dosage of mixed feed in organic farms, a lower concentration of thiamin and riboflavin in milk was established. Light contributes to decrease of the concentration of riboflavin in milk, therefore milking and pretreatment organisation in organic farms is one of the most significant factors, which impact the concentration of this vitamin.

The research results do not enable to raise the organic milk above any other milk obtained in other agricultural systems. Very often the organic farmers emphasise the wholesomeness and higher concentration of vitamins in organic products, however, the research results on the content of thiamin and riboflavin in organic milk do not prove it. It is very important hence the EU Regulation about food wholesomeness comes into force in Latvia. Any claim about higher concentration of vitamins and other substances in a product should be confirmed by research results. It is important for organic farmers too. The research results will help to make the accent and to popularise production of environmentally friendly and natural products.

Within the framework of research, the correlation between concentration of thiamin and riboflavin was proved. The correlation between the concentration of thiamin and riboflavin. The correlation coefficient $r>0.5$ (0.68) testifies, that there is a linear correlation between the concentration of thiamin and riboflavin. The p-value <0.05, it means, if the concentration of thiamin increases, the concentration of riboflavin increases as well, and vice versa. At the same time, this model cannot be used for forecasting of concentration of vitamins, because the determination coefficient is low ($R^2=0.47$), which is an evidence of a significant influence of other factors.

After evaluation of the chemical composition of organic milk, it was determined that the content of lactose and fat has significant difference ($p<0.05$) if compared with conventional milk. The concentration of calcium in milk samples, obtained from different agricultural systems, compared with data from literature, was significantly lower ($p<0.05$). The concentration of thiamin and riboflavin in organic milk was significantly lower ($p<0.05$), in comparison with conventional agriculture. Research results show, if the agricultural system converts from conventional to organic, the concentration of separate nutrients will convert too.

THE EVALUATION OF CONCENTRATION OF ANTIBODIES IN ORGANIC MILK

The concentration of immunoglobulins in milk is influenced by a variety of factors, including age, health status and stage of lactation. The concentration of IgA varies from 0.42 to 2.32 $g\ l^{-1}$ and it has significant differences ($p<0.05$)

in milks, obtained from different agricultural systems. The higher concentration of IgA was observed in conventional milk -1.12±0.07 g l^{-1}. The widest range of concentration of IgA was determined in milk, obtained from transitional period.

The concentration of IgA in milk samples decreases in the order: conventional ! transitional period ! organic agriculture. The mean concentration of IgA in organic milk was 5.5 times higher, in comparison with the data from literature (0.13 g l^{-1}). The concentration of IgA in milk obtained from transitional period and conventional agriculture was 6.9 and 8.6 times higher, relatively it could testify the mastitis problems in farms, where milk samples were obtained. The high concentration of IgA in conventional milk can relate with cows' vaccination, what is used in this agricultural system. The increased concentration of IgA was not related with mastitis, because milk samples were selected in such way, to eliminate the possibility to analyse milk samples obtained from mastitis cows. The conductivity of milk was detected and the conductivities parameters were used for selecting milk samples for analyses.

The dispersion analysis was used for comparing IgA concentration in milk samples, obtained from different agricultural systems. A significant difference ($p<0.05$) was determined in concentration of IgA in milk samples obtained from organic and conventional milk. The concentration of IgG in milk increases in order: conventional to transitional period and organic agriculture. The concentration of IgG (g l^{-1}) in milk samples obtained from agricultural systems was according to the data from literature - 0.15-0.80 g l^{-1}. The concentration of IgM has not significant difference in milk samples.

The mean concentration of IgM in organic milk was the highest -1.65±0.07 g l^{-1} and comparing with milk obtained from transitional period and conventional agriculture did not have the wide range results. The concentration of IgM increases in order: conventional ! transitional period ! organic agriculture. The concentration of IgM was significantly higher in organic milk ($p<0.05$). The mean concentration of IgM in milk samples from different agricultural systems was significantly higher ($p<0.05$) in comparison with the data from literature 0.04-1.00 g l^{-1}. Repeatedly increased IgM concentration in cows' blood could be explained as a reaction of inflammation. The concentration of IgM in blood and in milk is not proportional, therefore the increased concentration of IgM in milk could not be considered as the adverse body reaction caused by pathogens.

Increased concentration of immunoglobulins in milk could be evaluated as positive, because recently the interest for cows' immunisation with aim to maximise antibodies concentration in milk has increased. The concentration of lactoferrin was from 0.030 to 0.071 g l^{-1}. The concentration of lactoferrin increases in order: conventional ! transitional period ! organic agriculture. There is a significant difference ($p<0.05$) in the concentration of lactoferrin in milk from different agricultural systems. The concentration of lactoferrin in organic milk samples (0.03 g l^{-1}) was 1.5 times higher in comparison with conventional milk samples (0.02g l^{-1}). The concentration of lactoferrin in milk obtained from different agricultural systems was from 0.020 to 0.350 g l^{-1}, according to data from literature. The increased concentration of lactoferrin in organic milk was not related with mastitis, it was caused by other factors: cows keeping conditions, immunity boosting measures and others. Lactoferrin inhibits microorganisms, such as coliforms and others, for a high concentration of iron ions is required for existence and reproduction of these microorganisms. Thus a high concentration of lactoferrin in organic milk indicates the role of a potential antimicrobiological agent in raw material.

The concentration of lysozyme in milk samples obtained from different agricultural systems was in a wide range from 0.02 to 0.77 mg l^{-1}. The concentration of lysozyme is important parameter of milk quality; it impacts milk storage duration and milk processing. Lysozyme is termostable, 75% from lysozyme activity preserves after milk heating up to 75 °C 15 min or 80 °C 15 s. It means that in pasteurised milk the concentration of enzyme is sufficient for limiting development of microorganisms and for preserving milk quality. Although there was not found significant difference between organic and conventional milk samples, but there was marked tendency, the concentration of lysozyme increases in order: conventional ! transitional period ! organic agriculture. The research results show, that the concentration of lysozyme in organic milk is considerable, as antibody providing milk quality in due time.

The obtained results could be explained by different agricultural systems, where farmers used different methods for cow's immunisation. The cow's vaccination is used in conventional agriculture, but organic system is based on cow's natural immunity fortification without using medicaments. As a result, the concentration of antibodies in milk were different.

THE EVALUATION OF MICROBIOLOGICAL AND CHEMICAL CONTAMINATION OF ORGANIC MILK

The somatic cell count, total plate count, the concentration of cadmium, lead, copper, iron, zinc and the level of aflatoxin M_1 were determined in this research. Somatic cell count (SCC). Antibiotics and other medicine are forbidden in regular feedstuffs at organic farms, including antibiotic usage for mastitis treatment, disinfection agent usage is restricted too. All this measures can cause the decrease of somatic cell count and wide spectrum of mastitis agents in milk. The dispersion of somatic cell count in organic milk.

47.8% of organic milk samples fitted within the limits of Regulation (EEC) No 853/2004 of the European Parliament and of the Council laying down specific hygiene rules for the hygiene of foodstuffs. However 52.2% of milk samples showed the somatic cell count more than 400 000 mi^{-1}, including 25.0% SCC - above 1 000 000 ml^{-1}. The somatic cell count in organic milk ranged between 4

000 to 1 413 000 ml^{-1}, the mean SCC was 351 123±40 057 ml^{-1}, it was lower than in conventional milk samples at the same time period - 391 000 ml^{-1}.

Norman and Schaik consider, that somatic cell count in milk obtained from healthy cows is not more than 200 000 ml^{-1}, Konosonoka mentions in her research, that SCC in healthy cow's milk is 300 000 ml^{-1}. A higher somatic cell count is a measure for the prevalence of mastitis in a dairy herd and is used by regulatory agencies as an indicator of the wholesomeness, safety and suitability of raw milk for human consumption.

The main factor, which influences somatic cell count, is the state of cow's health. In organic agriculture the use of antibiotics for mastitis treatment is restricted. The restriction on antibiotics may lead positively to a more intensive health promotion and disease prevention efforts, and to consequent and early intervention in the case of disease. In this way it is possible to decrease number of antibiotic resistant bacteria in udder and in milk.

Furthermore, the cows' vaccination, using different vaccines, for example Staphylococcus aureus and others, increases. As a result some cause of mastitis is eliminated, but other multiply easy. Today there is not prevalent vaccine for mastitis prevention and in the near future will not be. In the nature nothing disappears and changes. Always the role of some microorganisms causing mastitis will replace and take other, therefore for mastitis treatment and prevention is necessary to look other ways, and resignation from antibiotic use is not considered as anomaly. Moreover, if this method will help to limit potential mastitis pathogens, it will be only positive solution.

On the other hand the restrictions on antibiotic use may negatively lead to a non-treatment policy, where animals requiring treatment will not be treated, as result the somatic cells count is higher than in this case. The staphylococci percentage of distribution in organic milk samples. From the obtained results it is possible to conclude, that milk samples with increased number of staphylococci (above 2 000 cfu ml^{-1}) can testify about cow's mastitis. Some of coagulase negative staphylococci (CNS) strains were identified for clarifying main cause of mastitis in organic farms. For isolates percentage, Staphylococcus aureus was isolated in 34% and CNS - 63% of analysed organic milk samples. From organic milk samples Staphylococcus strains were not isolated in 5 cases. Some of CNS strains were identified. Results have shown that coagulase negative staphylococci and Staphylococcus aureus have the leading role for initiate of mastitis in organic milk, what is related to Smolder's research results, where the main microorganisms prevailing in organic and conventional milk samples were CNS, respectively, 46.0 % and 40.9%, Staphyloccocus aureus, respectively, 15.8% and 18.6%.

Somatic cell count in milk is influenced by many factors. Bluzmanis considers that the increasing somatic cell count has a correlation with the amount of staphylococci. The correlation coefficient $r<0.5$ (0.204) testifies about weak linear correlation between somatic cell count and staphilococci count. B_0 coeficient's p - value (0.051) is higher than 0.05, therefore coefficient is equal to zero. It means that in this research correlation between SCC and the number of staphylococci was not found.

Not always the amount of staphylococci influences the increase of amount of somatic cell count in milk. SCC depends on the nature of mastitis causes, aggression and other factors. The other authors' researches verify that there is the correlation between Staphylococcus aureus and somatic cell count in milk. At the same time many CNS can cause clinical mastitis, without the rapid increase of somatic cell count. Therefore, for getting objective research results, it was necessary to look for correlation between pathogens and SCC. The correlation between somatic cell count and Staphylococcus species. The correlation coefficient $r<0.5$ (0.296) testifies about weak linear correlation between SCC and Staphylococcus species. The determination coefficient (R^2) is 0.088. It means that in 8.8% cases the changes of somatic cell count could be explained with the changes of amount of staphylococci depending on species of staphylococci. In this case it is not possible to find a correlation, but it is possible to describe the tendency.

p - value of coefficients (b_0 and b_1) was less than 0.05. With probability 95% it could be asserted, that b_0 and b_1 coefficients value is not zero, therefore tendency between SCC and staphylococcus species exists. With probability of 99% somatic cell count is significantly different in organic milk samples, where CNS and CNS and Staphylococcus aureus were isolated.

The highest increase of somatic cell count was in organic milk samples from which S.aureus and CNS were isolated. The highest count of staphylococci in organic milk, independently from staphylococci species, did not cause a significant increase of SCC. There is a weak linear correlation between the isolated microorganisms and SCC in milk. The total plate count is a parameter, which helps to consider the hygienic conditions in the farms. The total plate count in organic milk samples. Despite the high microbiological quality of organic milk, 27% of organic milk total plate count did not fit within the limits of Regulation (EEC) No 853/2004 of the European Parliament and of the Council laying down specific hygiene rules for the hygiene of foodstuffs - 100 000 cfu ml^{-1}.

In addition, for ensuring high quality milk, it is not enough to cool milk till +4+6°C in a short period of time (20 - 30 min). The initial quality of milk depends on: observing sanitary - hygienic rules during milking, cleanness of equipment and observation of personal hygienic rules. Only honest job and cleanness can make organic milk better than milk obtained from other agricultural systems. Aflatoxin M, (AFM_1). Of course, if there is a restricted usage of the preservatives for preparation of feed in organic agriculture, the risks linked to aflatoxins contamination could be higher in organic production than in conventional production. It appears that there are both advantages and disadvantages, regarding the aflatoxins problem in organic agriculture. Among the advantages included the use of less nitrogen in organic agriculture, and also the use of organic nitrogen. The disadvantages are that, since, no herbicides are used in organic farming, there are more weeds, as well as the presence of Pyralidae. However, in organic agriculture it is possible to reduce the presence of weeds and Pyralidae, by using crop rotations and other agronomic practices. The level of alflatoxin M_1 in organic and conventional milk samples.

The obtained results have shown that no AFM, values exceeded the permissible level - 0.05 $\mu g\ kg^{-1}$ in organic and conventional milk. The study relating to AFM, in milk samples found no significant differences ($p>0.05$) between organic and conventional milk. The research results relate with Wyss 's statement, that the level of alfatoxin M_1 has no significant difference in organic and conventional agriculture obtained milks. Martini established that 92% of organic and conventional milk samples were contaminated with aflatoxin M_1. Ghidini affirms that the contamination of organic milk with aflatoxin M_1 is significantly higher in comparison with conventional milk.

The research results show, it is possible to get silage with the level of aflatoxin M_1, which does not exceed the maximum levels for certain contaminants in foodstuffs set by Regulation (EC) No 466/2001, maximally decreasing usage of preservatives for feed preparation in organic agriculture. For producing safe milk for human consumption, the adopting preventive measures, as good agricultural, processing and storage practice, is necessary.

Heavy metals and trace elements. Organic methods are used to minimise pollution of air, soil and water, although they cannot ensure that products are completely free of residues, because of general environmental pollution, including acid rain, ozone hole and other. There are many of chemical hazards associated with foods. Contaminants in animal feeds, such as agricultural and industrial chemicals, heavy metals and other, can give rise to safety hazards in food of animal origin. However, organic agriculture does not reduce level of persistent environmental pollutants in organically grown products. These may therefore be present in organic feedstuffs and hence in organic food of animal origin. The concentration of lead and cadmium in milk samples, obtained from different agricultural systems.

Means values of lead concentration in organic and conventional milk samples were 0.025 ± 0.002 mg kg^{-1} wet weight and 0.033 ± 0.001 mg kg^{-1} wet weight. In both cases maximum permissible level for such product (0.020 mg kg^{-1} wet weight) determined by Regulation (EC) No 466/2001 setting maximum levels for certain contaminants in foodstuffs was exceed. The concentration of lead in organic and conventional milk have significant difference ($p<0.10$). The research results are in conflict with Ghidini conclusions, that there is no significant difference in the concentration of lead in organic and conventional milk. The main sources of lead pollution in the environment are: industrial production processes and their emissions, road traffic with leaded petrol, the smoke and dust emissions of gas-fired power stations, the laying of lead sheets by roofers as well as the use of paints and anti-rush agents. The lead contamination in organic and conventional milk samples could happen due to feeding the cows with fodder collected from along the road sides'. The cow acts as a very effective biological filter diverting lead from her feed to her bones rather than to her milk.

The mean concentration of cadmium in organic and conventional milk samples was 0.007 ± 0.001 mg kg^{-1} and 0.009 ± 0.001 mg kg^{-1} wet weight. The cadmium content in organic and conventional milk samples was very low and fairly constant in all types of milk. Cadmium, like lead, is a cumulative poison, i.e., the danger lies primarily in the regular consumption of foodstuffs with low contamination. However, in contrast with lead, the definition of an exact toxicity limit is not possible for cadmium. The decisive point is whether absorption of the existing cadmium actually takes place. This is, firstly, dependent upon the composition of the diet as a whole and, secondly, on the bio-availability of the cadmium compound present. Again, as with lead, the cow acts as an effective biological filter and the proportion of ingested cadmium finding access to milk is extremely small.

The concentration of iron, copper and zinc in milk samples, obtained from different agricultural systems. The concentration of iron in organic and conventional milk samples was 1.26 ± 0.02 mg kg^{-1} wet weight, what is for 0.04 mg kg^{-1} wet weight higher in comparison with conventional milk - 1.22 ± 0.03 mg kg^{-1} wet weight. The mean concentration of copper in organic milk was 0.210 ± 0.005 mg kg^{-1} wet weight, in conventional milk - 0.270 ± 0.007 mg kg^{-1} wet weight. The significant difference between organic and conventional milk samples was established ($p < 0.05$). The research results relate Bluzmanis results, where author found in blood serum, obtained from organic cows, in comparison with conventional, higher concentration of iron - 38.20 mg%, however lower concentration of copper - 0.06 mg%. There is not found significant difference between the concentration of zinc in organic and conventional milk, but the concentration of zinc in organic milk for 0.05 mg kg^{-1} wet weight was higher if compared with conventional milk samples. The research results relate to Bluzmanis research results, who determined the higher concentration of zinc in organic milk 0.32 mg%, in comparison with conventional - 0.19 mg%.

Some heavy metals (the so-called trace elements) are essential in very small concentrations for the survival of all life forms, for example, copper, iron, zinc and others. Despite this fact, it is often forgotten that in some circumstances, in higher concentrations, these can also be quite toxic, for example, when they are present in an organic compound.

The legally accepted upper limits for different heavy metals except lead though are not exceeded by any analysed milk sample, not even from conventional bulk milk. The obtained results verify, that separate parameters of milk chemical composition (lactose, fat, IgM and lactoferrin content) and quality is higher in organic milk samples, at the same time the content of thiamin, riboflavin and IgA is higher in conventional milk. This does not allow to highlight or to slight the quality of milk obtained from one or other agricultural system.

MILK QUALITY & SAFETY SERVE

The Milk Control section was founded about 55 years ago by the then Cape Town City Council after the Cape experienced a serious gastro – enteritis epidemic as a result of the distribution of contaminated milk. This service supports the different

Sub-Districts in the City of Cape Town by providing laboratory facilities where food, milk and soft serve samples can be tested and evaluated. It ensures the distribution of safe and healthy milk and dairy products by monitoring these products from the producer farms, the road tankers, through the pasteurizing and processing plants and finally to the retail outlets. It also works towards educating those who work with milk and dairy products to ensure high standards of handling.

LABORATORY SERVICE

The laboratory forms an integral part of the Milk Control section and its main function is to analyse all samples received. A service is also rendered to the various Sub-Districts where food samples from different outlets are tested. In order to determine the chemical composition of milk, the sample is tested for the percentage fat, protein, lactose and added water present. Bacteriological tests are performed in accordance with Regulation R1555 for milk and milk products.

Using prescribed methods and culture media the following tests are performed:

- Plate Count: to determine the number of colony forming units/1ml.
- Coliform Count: indicates the level of hygiene and cleaning. Modified Eijkman: indicates faecal contamination if Escherichia coli is present.
- Kundrat test: to establish antibioticsor antibiotic substances.
- Baira Parker Plate: to establish the presence of staphylococcus aureus.
- Aschaffenberg Mullen: indicates effective pateurization.
- Phosphatase test:

MILK CONTROL CORE BUSINESS

The core business of the Milk Control branch is to ensure the distribution of safe and healthy milk and dairy related products within the City of Cape

Town's boundaries by monitoring the product from the producer farms to the processing plants and bulk milk outlets.

MILK CONTROL FUNCTIONS

The Environmental Health Practitioners are responsible for the following functions:

- Producer farms are visited monthly and raw milk samples are taken for analysis.
- Hygiene practices are monitored strictly on farms through implementing health regulations relating to milking sheds and the transport of milk.
- Milk tankers are inspected regularly and loading procedures scrutinized.
- Dairy processing plants are monitored by sampling their products daily, weekly or monthly depending on the volume they produce.
- All plants are visited and inspected monthly to ensure acceptability of production procedures.
- Inspection reports are submitted to the responsible authorities.
- Bulk milk outlets are also monitored within the Unicity's boundaries – this is a new concept and arises from milk retailing in bulk to the public.
- Monitoring starts on the producer farm, the pasteurizing and distribution plant, transport to the depot and finally the milk as it sold to the consumer.

DEFINITION OF MILK

Milk is a fluid secreted by the mammary glands of healthy cows of the bovine species during the usual lactation period by means of complete and regular milking. It is a concentrated food, designed to initially protect the newly born animal by supplying concentrated antibodies to counter disease (passive immunity), until the immune system of the young animal is fully functional, and thereafter also to ensure rapid growth until the young are more independent.

The Composition of Milk

Next to blood, milk is the most complex of biological fluids and although the major constituents have been known for many years the list of minor constituents grows every year and at present more than 200 different constituents have been identified. The following table indicates the gross average composition of milk:-

Milk Composition, %
Fat, 3.8%
Total Protein, 3.4%

Casein, 2.7%
Albuman and Globulin, 0.7%
Lactose, 4.8%
Mineral Salts, 0.7%
Water, 87.3%

Vitamins

Although milk does not contain all the vitamins we require, it is a very important source of them.

The following are the more important ones found in milk:

- Vitamin A,
- Thiamine or Vit.B1,
- Riboflavin B2,
- Nicotinic Acid (Niacin),
- Vitamin B6,
- Vitamin B12,
- Vitamin C and
- Vitamin D.

Milk Products

Apart from the sale of milk, dairies have found a demand for a variety of other products made from milk. Low fat milk, skim milk and cream are commonly used, while yoghurts in the form of fruit yoghurt, plain and drinking yoghurt are growing. Other products are cheese, cultured buttermilk, ice cream, soft serve and milk desserts. All these products are tested by the City of Cape Town's Milk Control section for various health related risks such as bacterial contamination, phosphatase and the presence of antibiotics, as laid down by The Foodstuffs, Cosmetics and Disinfectants Act.

RAW MILK - A RISK TO PUBLIC HEALTH

Are you drinking raw milk, milk that has not been pasteurised to destroy disease-causing bacteria? Milk is a highly perishable food product and is an excellent medium for transmitting a variety of diseases.

How is Raw Milk Different than Pasteurised Milk?

Raw milk is unprocessed milk, which is consumed directly after milking the cow or goat. Pasteurised milk, on the other hand, has been heat - treated to destroy all disease causing bacteria that can be present in raw milk.

Have You been Told that Raw Milk is Better than Pasteurised Milk?

There are a number of myths about this. For example:

- Is raw milk better for you? No, there is absolutely no evidence to support the contention that raw milk is more nutritious than pasteurised milk.

- Does pasteurised milk contain chemical additives? No, pasteurised milk does not contain any chemical additives. In fact, law prohibits chemical additives.
- Is raw milk “pure”? No, raw milk can contain many disease-causing organisms. Raw milk for human consumption usually originates from unapproved and/or uninspected farms, which are not monitored for the presence of drug residues such as antibiotics.
- Does raw milk help keep you from getting sick? No, there is no scientific evidence that raw milk makes you more resistant to disease.
- Does raw milk help develop teeth better than pasteurised milk? No, there is no scientific evidence that raw milk promotes better development of teeth.

Some people may say they have grown up drinking raw milk and never suffered ill effects. However, public health authorities have documented many cases of illnesses caused by drinking raw milk, which tell a far different story. While mandatory pasteurisation of milk has virtually eliminated large outbreaks of milk borne disease in Canada, occasional cases still remind us of the hazards of raw milk consumption.

For example: Case studies done in Canada

- In Chilliwack, 5 members of a family became ill with Salmonella from raw goat’s milk.
- On Central Vancouver Island, 9 kindergarten children became ill with Campylobacter after drinking raw milk while on a school visit to a local farm.
- In Vernon, numerous cases of Campylobacter were traced to the consumption of raw milk.
- In the Kootenays, a 35 year old woman needed surgery and a long stay in hospital due to Brucella infection. She had previously consumed raw milk from several sources.
- In Ladysmith, a 4-year-old girl developed neck abscesses due to Yersinia infection. The family had consumed raw milk for many years. The older brother previously had a similar abscess in his neck and lip.
- On the Queen Charlotte Islands, 2 people developed Toxoplasmosis after consuming raw goat’s milk.

What can You do to Protect Yourself?

Buy pasteurised milk products from retail outlets. Milk that you buy from grocery stores is pasteurised and packaged in an approved dairy plant. Nobody is naturally immune to the disease - causing agents, which may be found in raw milk. However, infants and children are at more risk of getting sick from drinking unpasteurised milk because they usually drink a lot of milk, and are more vulnerable to infection.

Please Note:- For people living in rural areas who do not have access to commercially pasteurised milk the following process is recommended for home pasteurization:

- Disinfect milk bottles and tops by boiling in water for 10 minutes.
- Place raw milk in the top part of a double boiler. Gradually raise the temperature of the milk to 62 °C and hold this temperature for 30 minutes.
- Check the temperature of the milk frequently with a reliable thermometer.
- Cool the pasteurised milk promptly by placing the top part of the boiler in ice water. Stir frequently to speed up the cooling process.
- Pour the cooled milk into the disinfected containers and place under refrigeration at 4 °C.

Please also note:- Home pasteurisation by using a microwave oven is not a reliable method and should not be used as an alternative to the above process.

PASTEURISED MILK

The process of pasteurisation is named after Louis Pasteur, the French scientist (1822 - 1895), who became famous for his study of bacteria. In 1864 he discovered that spoilage could be prevented in wine and beer by raising the temperature for a short time. This process now called pasteurising, and is also applied to milk, which is an ideal medium for bacterial growth, including certain pathogenic types.

This is where pasteurisation is most beneficial and the temperature-time combination of the process is laid down by law Regulations Relating to Milk and Dairy Products - R 1555 of 1997 and Amendment R 759 of 2004, and is sufficient to destroy any pathogenic organisms which may have entered the milk and so that the authorities can be confident that public health will always be safeguarded. Pasteurisation does not sterilise milk, but will render milk safe, unless it is recontaminated after pasteurisation. The process of pasteurisation not only requires the heating of milk to a certain temperature for a certain time, but also requires immediate cooling to 4 °C and kept below 4 °C until sold.

Pasteurisation

The heating of milk in the pasteurising process nowadays is mostly the High Temperature Short Time method where the milk is pumped between closely spaced stainless steel plates causing a wide but thin film of milk. On the other sides of the plates, hot water flows, having been heated by steam and kept accurately at a temperature high enough to just bring the milk to its pasteurising temperature of 72 °C. The milk stays at this temperature for 15 seconds regulated by a Flow Controller. After this it passes through other stainless steel plates causing a thin film of milk as before and the heat of the milk is transferred to ice water flowing on the other sides of these plates until it reaches 4 °C and remain at this temperature until it is sold to the consumer.

Batch Method of Pasteurising

An alternative to the above system is a batch method where a tank of milk is heated to 63 °C and held for 30 minutes and then cooled. The heating and cooling is also done by water as above but in a jacket around the tank. Because of the long holding time, a cooked flavour is noticed in the milk.

Sterilised or UHT (Ultra Heat Treatment) Milk

The process of pasteurisation and batch method uses temperatures far below that of boiling water, which is 100 °C. However, if the milk were to be subjected to a temperature just above boiling water for a short time all bacteria would be destroyed. If it were kept sealed i.e. not recontaminated, it would never be soured by bacteria. On reopening it would of course become perishable like normal milk. This is sterilisation and is similar to preparing of a baby's bottle.

Tests Carried Out by the City of Cape Town

The City of Cape Town's Milk Control section will conduct a test called the phosphatase test to indicate if milk were pasteurised properly. All pasteurising plants inside the City and also those plants situated outside the City's boundaries distributing milk inside the City are subjected to these tests. Phosphatase is an enzyme present in milk and is inactivated by pasteurisation. The test is sensitive and one litre of raw milk is easily detectable in 1000 litre of pasteurised milk. The test result should be negative for phosphatase.

ASPECTS OF MILK RELATED DISEASE

Milk is a major factor in disease dissemination and also presents a favorable physical environment where microorganisms can multiply. Primary diseases in the cow play a minor part in milk-related disease in man.

This is especially true when one considers that the five major milk related conditions that originate from the cow are namely:

- Tuberculosis - Mycobacterium tuberculosis
- Brucellosis - Brucella abortus
- Anthrax - Bacillus antracis
- Food poisoning - Staphylococcus aureus
- Sore throat - Streptococcus pyogenes

Major milk-borne epidemics have all occurred as a result of the following:

- The environment - for example, contaminated water for washing surfaces and dust contamination.
- The withdrawing of milk from the udder may contaminate the product with a disease carried by the worker.
- Contamination by improperly maintained equipment.
- Storage and transport vessels may contaminate milk.
- Processing may lead to further degradation and contamination by ill-considered methods.
- Product handling, display and distribution to the final user could lead to an inferior product-for example the ease with which bacteria could multiply when temperatures rise to above 4 °C.

DISEASES ASSOCIATED WITH MILK

Brucellosis

All three species of Brucella have been isolated from milk in various parts of the world. In non-pregnant cows the organism is found in the secretory tissue of the udder. In man it causes systemic disease, usually with an insidious onset resembling flu, and clinical diagnosis is usually difficult.

Staphylococcus Infections

In man, pathogenic staphylococci are commonly associated with skin lesions, infected lacerations, boils and pustules that are common in farm workers. In cattle, staphylococcus is often found in older cows as a low grade, sub clinical mastitis or from small ulcers on the teats.

E-coli

E-coli is found naturally in the environment for example in soil, water and faeces/manure or in the digestive tract of humans and animals. It is an important indicator of poor hygiene practices or improper processing. Symptoms of contracting this organism are vomiting, fever, diarrhoea and stomach cramps.

Tuberculosis

Infected milk is the most important vehicle for the transmission of bovine tuberculosis in man, and it would appear that the presence in man is dependent on the prevalence in cattle, and the amount of raw milk consumed.

Streptococcal Infections

Group A streptococci (sore throat, scarlet fever and mastitis) originate from human carriers that in turn may affect the udder. Unpasteurised or inefficiently cooled milk can lead to a rapid multiplication.

Typhoid and Paratyphoid (Salmonellosis)

These are not natural pathogens of milk animals, but are readily transmitted by milk in those areas where pasteurization is not enforced. This is a common food poisoning organism due to poor hygiene or incorrect processing. Symptoms associated with contracting this organism are nausea, vomiting, abdominal pain, headache, chills and diarrhoea.

Shigellosis

Shigella infections in man are often associated with the ingestion of raw milk. It occurs worldwide and two thirds of the cases and most of the deaths are in children under the age of 10 years. This occurs commonly in population living under poor conditions-malnutrition, poor sanitation and crowding. Sources are hands of milkers, water and flies.

Campylobacter Jejuni

This organism is found naturally in soil, water, and farm waste and in the digestive tract of animals. In recent years there have been reports of severe outbreaks of enteritis in the UK attributed to Campylobacter jejuni. In all cases consumption of unpasteurised milk was implicated. Symptoms are profuse diarrhoea (sometimes bloody), stomach cramps, nausea, dizziness and fever.

Heliobacter Pylori

This organism may be transmitted through contaminated milk. Infection with the organism is a powerful predisposing factor to the development of stomach cancer. Treatment of clinical cases is expensive, and a correct diagnosis in humans requires a gastric biopsy.

Listeria Monocytogenes

This organism is found naturally in the environment. Consuming raw milk could lead to contracting this organism. It could cause symptoms like flu-like illnesses to meningitis. It may also cause abortion in pregnant women. It has a mortality rate of 30% of those who are infected.

MILKING PROCESS AND LEGAL REQUIREMENTS

The supply of milk to the consumer is a service, which often is taken for granted. Between the production of milk on the farm and its delivery to the consumer, a highly organised and efficient industry must function to ensure that consumer's demands are satisfied. The City of Cape Town' s Milk Control section is involved throughout this whole process in monitoring the product from the farm to the processing plants.

Milking and Milking Sheds

In South Africa it is seldom so cold that cows have to be kept stabled. Stables are used only to milk and feed cows. The most common method of milking is by machine instead of hand milking. The milk from the milking machines flows along a pipe to a refrigerated, insulated, stainless steel tank where it is chilled to = or <4 °C. These tanks have holding capacities of up to 5000 litres and are provided with stirrers ensuring a rapid cooling of the milk. Regardless of what method is used, milking is always preceded by the thorough washing of the udder to clean it of all contaminated material. The washing and sterilising of all equipment used and the cleaning of the stable is an important part of the stable routine.

Legal requirements:- These milking sheds are subjected to the provisions made in the Regulations relating to Milking Sheds and the Transport of Milk - R 1256 of 1986. In terms of these regulations, a milking shed must comply with all the provisions made and is deemed in all respects suitable for the production and hygiene handling of milk. The Local Authority will then issue Certificate of Acceptability if he is satisfied that the milking shed is suitable for production. The Environmental Health Officer in terms of this Act, has the authority to implement legal action should they not be in compliance with these regulations.

Transport of Milk from the Farms to the Factory

From the farm, the milk will be pumped from the refrigerated tank to a tanker which is an insulated stainless steel tank built onto a lorry. The driver of the truck, on arriving at the farm measures the volume of the milk in the tank by means of a graduated dipstick, he will then check the temperature of the milk, which should be 4 °C or lower and ensures that there are no foreign flavors present in the milk. The stirrers are then switched on and after the milk has been thoroughly mixed two samples are taken, one for bacteriological and the other for chemical analysis. If the driver is satisfied that the milk is sound, he then connects the farm tank to the tanker and pumps the milk out. The tanker will then continue to the next farm. When the tanker has a full load, it will proceed to the processing factory. The temperature should remain = or < 4 °C in this insulated tank.

Legal requirements:- The Minister of Health has promulgated Regulations governing Milking Sheds and the Transport of Milk, Regulation 1256 of 1986. In terms of these regulations the driver of the truck is required to comply with the Duties of a Driver as prescribed in the regulation.

The tanker should also be built to the specifications as required by the SABS CODE 1187 of 1978. Environmental Health Practitioners are responsible for inspecting these tankers regularly and have the authority to implement legal action if deemed necessary.

Reception at the Processing Factory

Even while being transported, the milk is kept at a temperature of = or < 4 °C, and upon its arrival at the processing factory the temperature is checked and the whole load of milk will be rejected if it is not cold enough. After this, further quality tests are done to ensure that bacterial activity has been kept to a minimum.

When the laboratory is satisfied with the quality, the milk will be pumped through stainless steel pipes to holding tanks (silo's), each of which in the larger dairies, is capable of accommodating 70 000 litres. After this, the tanker will be cleaned thoroughly. Facilities exist at larger factories for the internal washing of these tankers, which is done automatically by pumps, which pump water and detergents through spraying devices situated inside the tankers.

Legal requirements:- The Minister of Health has, in terms of section 35, read with section 40, of theHealth Act 63 of 1977, promulgated Regulations Governing General Hygiene Requirements for Food Premises and the Transport of Food - R 918 of 1999. In terms of these regulations, the person in charge, or the owner, is required to be in possession of a Certificate of Acceptability. The Environmental Health Practitioner will maintain a milk-sampling programme whereby random milk samples are taken and which are sent to the Laboratory for a bacteriological analyses. In terms of this Act they have the authority to implement legal action against any producer, manufacturer or supplier, should they not be in compliance with any section of this Act or any of the regulations made in terms of this Act.

Pasteurisation

From these first holding tanks the milk flows to the pasteurisers where it is given a heat treatment. The purpose of pasteurisation is to destroy any pathogenic organisms, which may have entered the milk. The heating of milk in the pasteurising process is mostly the High Temperature Short Time method where the milk is heated to 72 °C for 15 seconds and then immediately cooled down to 4 °C and kept at 4 °C until sold to the consumer.

Clarifying and Homogenising

A modern technique used to ensure that the milk is completely free of all foreign particles before it is pasteurised, and entails the continuous free flow of milk through a rapidly rotating centrifuge (7000 rpm). The clean milk flows out again in a pipe concentric with the inlet pipe. The homogenisation process is a method of breaking up each particle of fat until it is very small. Homogenising will make them all the same size as well as making the milk more palatable and digestible.

Filling and Washing

After the abovementioned processes, the pasteurised milk is held in tanks before being filled into containers. These containers are filled with milk by automated machines, which fill hundreds of bottles and sachets a minute. The containers are then put into crates and moved to the cold rooms to await delivery to the retail outlets.

It is of paramount importance that each piece of equipment is thoroughly washed each day after use. Pipelines and pumps used for milk are designed in circuits so that, for cleaning purposes, detergents can be pumped around internally.

Delivery

Some dairies dispatch straight to the consumer while some dispatch to local distribution points, cafes and supermarkets. It is very important that the milk's temperature will remain at = or < 4°C during this delivery process. We know from experience that milk can turn sour and the reason for this is mainly due to the bacteria, which turn the milk sugars to acids. These bacteria are mostly active at warm temperatures, and at low temperatures this activity is greatly reduced. Because milk is also a good nutrient for bacteria it must be kept very clean and cold and this is what makes the whole operation very difficult.

MICRO-ORGANISMS RELATED TO MILK

The classification of the various types of bacteria according to the type of reaction they could cause in milk, or according to their temperature related characteristics, is very significant from the milk industry's point of view. Important bacterial groups, according to the abovementioned classification, are as follows:-

Acid Forming Bacteria

These bacteria will cause a sour taste in milk and will lead to a pH drop in milk from 6,6 to a pH of 4,6. A pH lower than 5,2 will coagulate the casien, which causes the milk to thicken. Acid formers are mainly mesophilic bacteria and will dominate in milk that has not been cooled properly.

Gas Forming Bacteria

Lactose fermentation by these bacteria causes a considerate amount of gas forming, namely carbon dioxide and hydrogen gasses.Examples: Coliform bacteria causing early bloating in cheese Clostridium butyricum - a spore forming bacteria.

Sweet Curdlers and Ropiness

Bacteria causing these problems in milk are able to increase the viscosity of milk without causing considerable acid forming.Examples: Alcaligenes viscosus- from stagnant water and Bacillus subtitles- from dust

Proteolytic Bacteria

This type of bacteria will decompose casein, causing faecal odors and a bitter taste in milk. Due to the putrefaction of the protein, the milk becomes watery in appearance. Example: Clostridium batyricum

Psychrotrophic Bacteria

These bacteria are able to multiply at temperatures as low as 2 °C - 10 °C. They are the most dominant bacterial population in milk that is stored for two days or longer at 10 °C or lower. They cause a bitter taste in pasteurised milk. Examples: Pseudomonas species

Mesophilic Bacteria

Mesophilic bacteria can multiply at room temperature between 5 °C and 45 °C, but have an optimum growth temperature of 25 °C to 37 °C. These bacteria are the most dominant in nature. It is therefore very important to keep milk below 4 °C. Examples: Streptococci lactis - acid forming in milk and Micrococci pyogenes - causes mastitis

Thermophilic Bacteria

These bacteria can often survive the pasteurisation process. Examples of temperature and time combinations for pasteurisation include: 72 °C for 15 seconds or 62,8 °C for 30 minutes Examples: Micrococcus Streptococcus

Spore Forming Bacteria

Spore forming bacteria are able to form endospores, which will survive a heat treatment of 80 °C for 10 minutes. Examples: Bacillus cereus, Bacillus subtilis and Clostridium butiricum

Coliform Bacteria

The presence of these bacteria requires urgent attention for the following reasons:

- Their relationship with faecal contamination (E-coli)
- The 'off ' tastes they cause in milk Coliform bacteria are an indication of good sanitation of all milk contact surfaces and equipment.

When coliforms are present in pasteurised products, it almost always indicates that spoilage bacteria are present. Example: E-coli (faecal contamination)

MILK FLAVOUR QUALITY CONTROL

FLAVOUR QUALITY CONTROL

Some of the more common off-flavors that may be present in milk as it comes from the cow or may develop during processing and storage were discussed in the first two papers in this series. Understanding the nature of desirable and undesirable flavors in milk, and conditions that influence these flavors, provides a sound basis for an effective flavour control programme. In addition, not only quality control personnel, but also milk producers and processors, can benefit materially by familiarising themselves with techniques for flavour perception and evaluation, and by developing proficiency in identifying the common off-flavors of milk.

Preparing Milk Samples

The first important step in a flavour control programme is the technique for collecting and preparing milk samples. Commercially processed milk in its consumer package usually serves as a satisfactory sample and container. The products must be protected from light, refrigerated during transport, and either examined promptly or stored at a standardised temperature before examination. Milk samples that must be collected from bulk containers or individual cows pose special problems. Frequently, the method of treating the sample must be adapted to the flavour defect that is being studied.

For most purposes, milk bottles (usually a quart) are satisfactory and conveniently available. Bottles with rubber closures should be avoided as the rubber may impart foreign flavors to the milk. Polyethylene bottles with polyethylene screw caps are very satisfactory. They are light and durable, withstand heat shocks of laboratory pasteurization, and may be autoclaved if desired. For studies of oxidized flavour, they have an additional advantage over glass they adsorb less copper that would otherwise contaminate the samples.

Raw milk should be pasteurised before it is tasted, but the time and method depend on the defect. For many purposes, the milk should be pasteurised immediately and examined while fresh and again after a standardised storage treatment. Pasteurization, however, prevents the development of rancidity by inactivating the lipase. If rancidity is of concern, the milk should be stored raw for at least two days and then pasteurised before the flavour examination. on the other hand, if the greatest concern is with oxidized flavour, the milk should be pasteurised before storage to prevent the development of rancidity and bacterial flavors that might otherwise mask oxidized flavour.

Pasteurization of flavour samples in accordance with legal definitions is not practicable. Samples may be pasteurised in their containers by heating and cooling them in a water bath, with a thermometer in a reference bottle containing the same volume of milk as the sample bottles. Shaking the bottles provides adequate agitation. The containers should be closed to minimize volatilization of feed flavors. The temperature-time combination used should produce little or no heated flavour, yet provide protection from disease and be convenient for the equipment available. If a thermostatically controlled agitated water bath is used, heating to 145 F for 30 min. is satisfactory; otherwise a combination such as 150 F for 4 min. or 161 F with no holding period is more convenient. It is difficult to recognise flavors from descriptions. Hence, it is important to practice identifying the most common defects by repeatedly tasting samples with known flavors. The procedures outlined below are suggested as convenient methods for preparing samples having typical flavour defects.

Feed or Weed flavour

Where facilities permit, it is best to feed an individual cow feeds or weeds known to cause characteristic flavors (e.g. silage) 1 to 2 hr. before milking, then collect the milk from this cow. In processing plants it is usually possible to select milk with feed flavour from individual producers during receiving operations. Samples prepared by soaking feeds in milk do not usually give typical defects.

Oxidized flavour

Place a strip of bare sheet copper or 1 ft. of coiled copper wire in a bottle of pasteurised milk (not homogenised) and store in refrigerator one or two days. Unless the milk is unusually resistant, oxidized flavour will develop.

Light-activated Flavour

Light-activated flavour is usually evident in freshly processed homogenised milk exposed to sunlight for 2 hr. in quart glass bottles, and then held in a refrigerator for at least 4 hr. Homogenised milk is preferred because it is more susceptible]e to light activated flavour and less susceptible to oxidized flavour than non- homogenized milk. Adding ascorbic acid (50 mg./quart) before exposure enhances light-activated flavour but inhibits oxidized flavour.

Rancid Flavour

Mix equal volumes of raw milk and homogenised milk and store in refrigerator overnight. If rancidity is not evident, heat to 70-90 F and hold. Taste at hourly intervals until the desired intensity develops, then pasteurised rise.

Salty Flavour

The best way to obtain a typical salty flavour in milk is to examine samples from individual cows and select one with the defect. Salty flavored milk is produced most frequently in advanced lactation or when the cow has mastitis.

Heated flavorP> Heat milk to a temperature higher than is normally used for pasteurization, such as 170 F for two min., and cool.

EVALUATING MILK FLAVOUR

Understanding how we perceive flavors aids in flavour evaluation. Flavour is a blend of sensations from several sensory organs. The most important sensations contributing to milk's flavour are the taste, smell and feel. Taste is detected in the mouth, principally on the tongue. There are only four tastes—sweet, sour, salt and bitter. To produce a taste sensation, a substance must be in solution in order to diffuse into the receptor sites in the taste buds.

Odour is detected by olfactory nerves in the nasal passages. The number of odors is almost limitless, and many are difficult to describe. To have an odour a substance must be volatile. Most compounds with strong odors are more soluble in fat than in water, whereas the reverse is true of most substances with pronounced tastes. Another sensation that contributes to flavour is the way the food feels in the mouth. Some tactile characteristics of milk are smoothness, chalkiness, and richness attributable to consistency.

Although most milk is drunk cold, warming milk enhances its odour and facilitates detection of off-flavors. For quality control evaluations, milk usually warmed to about 70 F. The bottle should be shaken, the cap removed and the sample smelled. A head space above the milk (e.g. bottle two-thirds full) aids in evaluating odors.

Next, the sample should be tasted. The milk should be moved around in the mouth to contact all portions, particularly of the tongue, and to promote volatilisation of compounds that contribute to the odour. Particular attention should be given to the changing sensations with time. Sweet, sour and salty tastes are usually detected quickly, whereas bitter tastes become evident more slowly but are retained longer.

The most important step in learning to evaluate milk flavour is to gain proficiency in identifying defects. Correct identification provides helpful guidelines for control of defects. Evaluating the intensity of defects, however, is also of value in quality control programmes. Many different systems for designating intensity of flavors are in use.

In some, intensity is indicated only by descriptive terms such as slight, distinct or strong, or by one, two or three plus signs. A simple numerical system applies scores for these descriptive terms: 0, no criticism; 1, slight; 2, distinct; and 3, strong. An inverse of this scale, or one giving a higher maximum score for flavour, such as 5 or 10 points, is preferable for quality scoring as it recognises higher quality with a higher score. Any numerical system is arbitrary, but there is an advantage in selecting one that is in common use. In the United States a system that is widely used assigns scores within the range 10 for good milk to 0 for poor milk.

TROUBLE-SHOOTING FLAVOUR DEFECTS

If an off-flavour is found in a milk sample, a systematic approach helps in identifying the defect and its cause. An experienced individual will identify the most common off-flavors of milk by taste, and will be able to proceed immediately to determine the most probable cause. A beginner in flavour quality control work will be guided in his identification of off-flavors by comparing defective samples with samples having known off-flavors. Changes in intensity of a defect during storage provide helpful evidence regarding its identity. Hence, samples should be tasted fresh and again after storage in a refrigerator for at least 48 hr.

RAW MILK

To pin-point possible causes of off-flavors in raw milk, it is helpful to collect milk samples at different steps. Samples might be collected from individual cows, at the discharge from the pipeline milker, from individual cans or the bulk tank, from the tank after only one milking and later at the time the milk is collected, or from morning and evening milkings separately. If an off-flavour is present in a fresh sample collected from pooled raw milk, it is usually attributable to feed. In some cases, there may be a marked difference between morning and evening milk in intensity of the off-flavour. Examining the feeds for known troublesome materials may substantiate suspicions regarding possible causes.

Recommendations for corrective measures include:

- Use only feeds that cause little or no feed flavour. Eliminate weeds from pastures and from crops to be used for hay or concentrates.
- Arrange feeding schedule to
- Prevent the cows from eating feeds that may cause off-flavors during the 4-5 hr. before milking. Provide an environment where the cow can breathe fresh air free of feedy, cowy or barny odors.

Individual cows in a herd may produce milk with an off-flavour when it is drawn, such as a salty taste attributable to mastitis or advanced lactation, or a cowy flavour caused by ketosis. It is rare that such defects can be identified in the mixed milk from the entire herd, but they make the pooled supply less palatable. Therefore, milk from such cows should be withheld from the pooled milk.

In some instances, the cow may appear to be responsible for off-flavors that are actually caused by equipment, or treatment of the cow or equipment. Examples are a medicinal flavour from salves used on the teats or udder, or chlorophenol flavour from plastic or rubber parts of the milking equipment.

If an off-flavour is not present in fresh milk, but develops during storage, possible identities are light activated, oxidized, rancid and microbial flavors. Light-activated is the least likely in raw milk, and in any case its cause and correction would be immediately apparent. Comparison with reference samples aids in differentiating between oxidized and rancid flavors. Also, pasteurization of the fresh sample prevents development of rancid and bacterial flavors but not oxidized flavour. Some microbial flavors caused by psychrophilic bacteria are difficult to differentiate from rancid flavour because the bacteria produce lipolytic enzymes that release fatty acids, and also proteolytic enzymes that yield degradation products from protein with similar flavors. Information from bacterial counts of the milk, inspection of equipment, and checking on sanitising procedures and cooling practices indicates whether bacterial flavors may be involved. If the defect appears to be rancid flavour, a chemical test for free fatty acids could be used for confirmation.

Rancidity that occurs in raw milk supplies usually is induced by an activating treatment: either excessive agitation of warm milk or temperature fluctuations between 50 and 86 F. Eliminating the activating condition usually prevents development of the defect. Determining the susceptibility of milk from individual cows is also helpful. Cows that produce the most susceptible milk are usually in advanced lactation, and there is little loss in production resulting from drying them off.

In correcting problems with oxidized flavour. If samples taken from individual cows indicate that a high proportion of the cows in the herd are producing milk in which the defect develops without metal contamination, the most practical control would be through the herd ration.

PASTEURIZED MILK

In 'trouble-shooting' causes of defects in processed milk, samples should be collected from the raw milk storage tanks and at every step in processing where feasible. A good practice is to compare 'first-off' samples with samples collected near the end of a processing run. If product from one processing line feeds several fillers, samples should be taken from each filler. As for raw milk, the samples should be tasted fresh and after storage, except that a storage period of at least a week is preferable for most pasteurized products.

Identification of the defect, using prepared reference samples for comparison if necessary, is the first step towards effective control. Corrective measures for the most common defects were indicated in the first two papers in this series, and are summarised briefly here.

Limited shelf-life resulting from bacterial growth is usually caused by post-pasteurization contamination. Phosphatase tests may be run on freshly pasteurized products to check on adequacy of pasteurization. The unclean, fruity, bitter and putrid flavors that develop in pasteurized milk are usually attributable to growth of psychrophilic bacteria that do not survive pasteurization. Hence, critical evaluation of cleaning and sanitising procedures is necessary in order to detect sources of post-pasteurization contamination. Checks on temperature and time of storage, and rotation of stock should also be included. Rancid flavour should not develop in a properly pasteurized product, but activating treatments at some stage of processing may induce its development before pasteurization. Possible causes include mixing pasteurized homogenised products with raw milk, warming cooled milk to about 85 F and recooling, and excessive agitation and foaming of raw milk. If the intensity of the flavour increases during storage, the possibility that the milk was improperly pasteurized or was contaminated by raw milk should be checked by use of the phosphatase test.

Problems with oxidized flavour may be caused primarily by low oxidative stability of the milk as produced, but are aggravated by abuse at any stage of processing and distribution. Particular attention should be given to avoiding copper contamination from white metal and other copper-containing alloys, and to minimizing exposure to light. Use of higher pasteurizing temperatures may be helpful if the more severe heat treatment does not produce objectionable heated flavors. Milk that is susceptible to oxidized flavour may be directed into homogenized products, as homogenization inhibits oxidized flavour. Locating causes of some atypical flavors in processed milk is sometimes a baffling assignment, even for people with extensive experience in flavour-control work. However, provide guides to step-by-step elimination of possible causes of a defect.

MILK FLAVORS

Flavour in one of the most important qualities that determine the acceptability of milk. Even though milk is highly nutritious, people will not drink it if they do not like it. Hence, milk should be produced under conditions that give good flavour initially, and also be handled to protect its flavour at every step from the cow to the consumer.

As population densities increase around large metropolitan centers, milk is usually transported longer distances to supply the urban population, and the farms located near the cities tend to increase their scale of operations. Both changes may influence milk's initial flavour and particularly its flavour stability. Transporting milk longer distances usually increases the time between production and processing, which provides a greater opportunity for development of off-flavors.

To maintain bacteriological quality during this extended storage period, greater emphasis is placed on improved sanitary practices and effective refrigeration. Defects of bacterial origin are generally kept under control, but the changes may result in an increased incidence of other defects. For example, as dairying became more intensive, bulk collection systems were introduced. They provided for cooling milk to lower storage temperatures, and when couple with longer storage times, oxidized and rancid flavors could develop. When the average number of cows in herds increases, farmers have to provide a larger proportion of the cow's feed in dry form as hay and concentrate, instead of as pasture and silage. In general, dry feeds yield milk with greater susceptibility to oxidized and rancid flavors than do succulent feeds. Feeding high levels of concentrate, as practiced in intensive dairy areas, increases the concentration of unsaturated fatty acids in milk lipids, with an accompanying increase in liability of the fat to oxidation.

As distribution channels are equipped to provide more effective refrigeration of milk during delivery to stores and homes, and during storage in retail outlets, processors extend the expected shelf-life of their products. For example, a Californian company dates the products it packages in paper containers for expected 'shelf-life' of 12 days in distribution channels, and extends its tests in the quality control laboratory for an additional week. Thus, it is not sufficient to produce milk with good flavour initially. Increasing attention is being given to the stability of milk flavour during storage.

FLAVORED MILK

Flavored milk is a sweetened dairy drink made with milk, sugar, colorings and artificial or natural flavorings. Flavored milk is often pasteurized using ultra-high-temperature (UHT) treatment, which gives it a longer shelf-life than plain milk. Pre-mixed flavored milk is sold in the refrigerated dairy case alongside other milk products. Flavored sweetened powders or syrups which are added to plain milk are also available.

Types

Flavored milk is milk that has sugar, colorings and (mostly inexpensive artificial) flavorings added to make it more appetizing, especially to children. It can be sold as a powder to be added to plain milk, or bought pre-mixed alongside other milk products. Flavoring can be included in a straw, and some flavored milk products are designed as dietary supplements by including additional vitamins or minerals.

Bottled spiced (masala) milk is a popular beverage in the Indian subcontinent. Other companies provide flavored beverages in the United Kingdom, which sells packaged beverages to the mobile vendor market. Australia has the highest consumption rate of flavored milk in the world, standing at 9.5 liters per capita in 2004.

Flavored milk is particularly popular in the Australian states of South Australia and Western Australia. A 2013 Sunday Times article reported Western Australia was the "flavoured milk capital of Australia", with a A$220-million industry, average consumption of 19 liters (5.0 U.S. gal) per person, and more than 40 varieties of iced coffee alone available. Similarly, a 2006 Adelaide Advertiser reported South Australia consumed 45,000,000 L (12,000,000 US gal) of flavored milk each year, with 82% of market share held by a single brand,

Farmers Union. According to Coca-Cola Amatil, one of the largest bottlers in the Asia-Pacific region, South Australia is the only place where sales of flavored milk outstrip those of cola.

Controversy and Criticism

Jamie Oliver, host of Jamie Oliver's Food Revolution, brought attention in the United States to public schools that serve flavored milk in the school cafeterias. Flavored milk advocates claim that many children will avoid the nutrition found in milk unless it has been flavored, with the benefits of milk outweighing a few teaspoons of sugar. Opponents say that with rising levels of obesity and heart disease, flavored milk should be removed from schools and children should be taught to drink regular milk.

NORMAL MILK FLAVOUR

Milk of good quality is a very bland food with a slightly sweet taste, very little odour, and a smooth, rich feel in the mouth. Because of its bland flavour, the presence of minute quantities of abnormal constituents frequently results in off-flavors. Dairy producers and processors have, of necessity, been so concerned with the control of off-flavors, that until recently little attention has been given lower temperatures than prevailed with can collection systems. Most people associate the palatability of milk with its 'richness'. It is generally assumed that milk fat is one of the most important constituents in contributing to the desirable flavour of milk. Although there is a close correlation between the concentration of fat and solids- not-fat in milk as produced by the cow, little attention has been given to the contribution of the solids-not- fat constituents to its flavour.

In studies conducted in the 1960s, tests were made to determine the minimum difference in the concentration of selected milk constituents that could be detected reasonably consistently (two times out of three) by a trained laboratory panel. Special care was taken in the preparation of the milk samples to avoid any abnormal flavors that might have influenced detection of differences. In homogenised milk of typical commercial composition (3.5 % fat and 8.0 % S.N.F.), the percentage of fat had to be increased by more than S.N.F (i.e. to over 9 %) before the panel could differentiate it from the control milk two times out of three, whereas the S.N.F. had to be increased by only 0.6% for comparable differentiation. This result indicates that a small increase in S.N.F. would have a greater influence on milk flavour than a comparable increase in fat.

The experiments were extended to study the roles of individual constituents or fractions of S.N.F. An increase of 0.33% lactose was detected two times out of three. A concentrate of non-dialysable constituents of skim milk (mainly protein) was prepared by dialyzing concentrated skim milk against normal skim milk, and this concentrate was used to increase the concentration of protein and other colloidal constituents in milk. An increase of 2.S% of non-dialysable constituents (2.2% of the increase as proteins) was needed for two-thirds correct detection. Thus, lactose appears to be important, and the proteins relatively unimportant, in contributing to the normal flavour of milk.

The increasing sales in the United States and Canada of milk products containing increased S.N.F. concentrations indicate that these products have acceptable flavour. Products containing 2% fat and 10% S.N.F. are widely distributed and are gaining in popularity. Skim milk with a normal S.N.F. concentration has never been popular, but non fat products with increased S.N.F. have developed into successful commercial products.

Cows' milk is a palatable beverage, but nature did not develop it to appeal to man's taste. We should not limit ourselves to using milk as nature produces it for the calf. To increase milk sales, we should take advantage of modern technology to modify milk to man's taste and nutritional requirements.

OFF-FLAVORS IN MILK AS PRODUCED

Feed and Weed Flavors

In many countries the most common flavour defect of milk is feed flavour. The incidence in different countries is hard to assess because evaluations are based only on subjective judgements, and opinions regarding the intensity of the off-flavour that constitutes a defect are very variable. Likewise, levels that are objectionable to consumers are equally variable. The presence of feed or weed flavors in a high proportion of milk samples in a number of surveys indicates that the off-flavors may be detected by, and are presumably objectionable to, many consumers.

Understanding the mode of transmission of flavour substances in the cow's body assists practical control of feed flavors. The respiratory system and the digestive tract are both important in transmitting flavour substances to the milk odors from some feeds pass into the cow's blood from air in the lungs, and are then carried to the udder and appears. in the milk. Some flavour-producing substances are absorbed by the blood from the digestive tract, and are then transmitted to the udder. For some feeds, both the respiratory and digestive tracts are involved in the transmission of flavour to milk. Some feeds, such as garlic and onion, release volatile flavors after partial digestion in the rumen. odors belched from the rumen are inhaled into the lungs and transferred to the blood. This pathway provides a more rapid transfer of feed flavors from ingested feed than direct absorption from the digestive tract.

Fortunately, blood provides a two-way street for transportation of feed flavors. When the concentration of the flavour substances is higher in the milk in the udder than in the blood, the substances transfer from the milk to the blood. If sufficient time is allowed after the feed is consumed, the flavour substances are eliminated from the blood, partly by transfer of volatile substances to the air in the lung, and partly by metabolism of the substances. In either case, they are eliminated from the cow's body.

The time interval between eating and milking is an important factor influencing the intensity of feed flavors. Whether the off-flavour resulted from only breathing the odour of silage, or from eating the silage, the feed flavour was most pronounced from 2 to 3 hr. later, but had been eliminated from the milk 5 hr. later. These results show why feeds commonly used in dairy rations will not cause objectionable flavors if they are fed after milking, and withheld from the cows during the 4-5 hr. before milking. Not all feeds respond to flavour control. Some flavour substances accumulate in cow body tissues, particularly in the fat. They then transfer to the blood, and hence to the milk, over long periods of time. one of the most lingering flavour defects reported resulted from eating potatoes grown on a field treated with benzene hexachloride (for nematode control). A mothball-like flavour was still detectable in the milk three weeks after this feed was discontinued.

Flavors from some weeds persist for longer than 12 hr. after they are eaten, and therefore such weeds must be kept out of a cow's ration. In selecting feeds for a dairy ration, one criterion must be that either the feed does not impart an undesirable flavour to milk, or the flavour can be controlled by withholding the feed for a reasonable time before milking. It is impracticable to withhold feed for more than 5 hr., as prod production will suffer. Flavors from many feeds can be controlled by withholding the feed 2-4 hr. before milking.

Reports on some feeds and their effect on milk flavour are inconsistent. For example, sugar beets, beet pulp, and beet tops do not normally cause objectionable flavors, but are sometimes responsible for fishy flavour. The difficulty appears to be sporadic, varying with each individual cow and with different lots of feed. The quantity of feed consumed, its quality, and in some cases its maturity, may influence its flavour- producing characteristics.

Many weeds cause off-flavors in milk. Weeds are variable in their effects. Flavors caused by some can be controlled if the weeds are withheld from the cow 5 hr. before milking. The off-flavour from some weeds (onion, penny cress) appears principally in the milk fat, whereas that from others (bitterweed) is associated with the skim milk portion. The defect may be more pronounced with young plants (cocklebur) or with older plants or their seeds (penny cress). In most cases the off-flavour is present when milk is drawn from the cow, but this is not always true.

For example, skunkweed flavour does not appear in fresh milk or cream, but shows up in butter or cheese. Wild carrot eaten alone sometimes produces a mild off- flavour, but when eaten along with mule's tail or mare's tail, the skunkweed defect becomes pronounced. The scorched flavour caused by swine cress is evident only after milk or cream is heated. on the other hand, pasteurization markedly reduces or changes the character of other feed flavors. The ensiling process destroys the substances causing ragweed and pepper grass flavors but does not affect the substances causing onion flavour. Such inconsistencies in the influence of different feeds and weeds on flavour indicate the danger of making general statements regarding their behaviour.

Since, odoriferous substances are volatile, attention has been given to developing practical procedures for removing them by vaporization. The value of using aerators or surface coolers as a means of reducing feed flavors is often overrated. When alfalfa-flavored milk cooled over a surface cooler was compared with the same milk cooled in a container, experienced judges had difficulty in detecting consistently the sample with supposedly less feed flavour. Although aeration may cause some reduction in the intensity of some feed flavors, it cannot be relied on to control the defect.

In recent years many liquid milk processors in the United States and Canada have installed equipment that subjects milk to a combined vacuum and heat treatment to remove volatile flavors. The effectiveness of such equipment depends not only on the intensity of the treatment, but also on the nature and concentration of the flavour substances. Some flavour treating machines are very effective, whereas others have little influence on milk. Use of flavour-removing equipment can provide the processor with an extra safeguard in protecting the flavour quality of the milk he sells, but present indications are that the producer will still have to take the major responsibility for control of feed flavors by sound management practices.

Cowy, Barny and Unclean Favors

These terms describe flavors that are attributable to unsatisfactory production conditions, but there are several possible causes. one of the most common is inhalation by the cows of foul air in poorly-ventilated barns or corrals in which wet manure has accumulated. The barny, unclean odors are transferred to the milk through the cow's respiratory system in the same manner as for feed flavors. Although direct absorption of the odors by milk during and after milking is frequently mentioned as a possible cause, research has established that this source is insignificant compared with absorption through the cow's respiratory tract.

Metabolic disturbances of the cow may also result in cowy and other unclean flavors. Ketosis is frequently accompanied by cowy flavour. Cows with the disease have a high concentration of acetone bodies in their blood. These compounds appear in the milk and are at least partly responsible for the cowy flavour. Other conditions that upset the cow's d digestive processes have also been associated with cowy and other undesirable flavors.

Mastitis reduces the flavour quality of milk, but the effects are variable. Mild cases may result in a flat flavour, probably due to a lower concentration of the normal milk constituents. In more severe cases of mastitis, the milk is usually criticised for cowy, unclean, and salty flavors. In some cases cowy, barny and unclean flavors may be attributable to the cow's feed. Some weeds cause the appearance in milk of indole and skatole compounds that give the characteristic odour to faces. Abnormal silage fermentations may also result in unclean flavors in milk.

Although cowy flavour is listed as a defect present in milk as drawn from the cow, the flavour may intensify during storage. The increased intensity may be due to increased concentrations of acetone and butanone, but nothing has been reported regarding the source of and conditions influencing the appearance of these compounds during storage.

Foreign and Chemical Flavors

Some flavors that are present immediately after milking may be caused by chemical contamination. Treatment of teats with salve is a possible source of medicinal flavour. Equipment used in handling milk may also contribute off-flavors, such as the phenolic taste from certain plastics, or chemical tastes from sanitizing agents used to treat milking equipment. Combinations of chemical contaminants may be more troublesome than one alone, as illustrated by difficulties encountered by some dairy producers and processors with chlorophenol flavour. Phenolic and chlorine compounds in combination can, in certain circumstances, be detected at lower concentrations than either alone, because they may react to form chlorophenols.

Salty Taste

Salty taste is caused by two conditions: mastitis and advanced lactation. The defect is rarely detected in pooled milks, because only a small proportion of cows would produce salty milk in a well-managed herd. Nevertheless, salty milk should be eliminated from the supply, because it is detrimental to milk quality.

Milk that has excellent flavour as produced may develop objectionable defects before it reaches the consumer. Recent changes in the dairy industry have caused changes in the flavour problems encountered. In the early days of the market milk industry, when sanitation practices were inadequate and relatively few refrigerators were available in homes, defects caused by bacteria, such as sour flavour, limited the time that milk remained usable.

Today such bacterial defects are encountered less frequently because of modern sanitation practices, improved refrigeration and transportation facilities, and widespread adoption of pasteurization. Defects caused by chemical changes during storage such as oxidized and rancid flavors make up a larger percentage of observed defects than formerly. Milk as purchased is now more uniform in flavour and defective product is more likely to result in a complaint than in former years. Hence, processors must pay more attention to improving both the initial flavour and the flavour stability of milk during its expected usable life.

Oxidized Flavors

Oxidized flavour is a troublesome defect of non-homogenised milk, skim milk, cream, and certain other dairy products. The flavour is described by terms so various as metallic, papery, cardboardy, oily, and tallowy, indicating the great variability of the predominant flavour characteristic. The defect is caused by the oxidation of fatty constituents in the product. Knowledge of the milk constituents that are involved in the development of oxidized flavour, or influence susceptibility of milk to the defect, helps one to understand the factors that influence the flavour. The compounds responsible for the flavour are produced by oxidation of unsaturated fatty acids in the phospholipids in the membrane surrounding the fat globules. Conditions are more conducive to oxidation at the surface than inside the fat globules for at least four reasons: (1) The phospholipids, which are concentrated at the surface lace, contain a higher proportion of unsaturated fatty acids than the remainder of the fat globule; (2) The oxidation catalyst, copper, is also concentrated at the surface of the fat globule; (3) oxygen from air dissolves in the skim milk portion of milk, and thereby reaches the surface of the fat globules; (4) The ascorbic acid in milk, which is also dissolved in the skim milk portion, is an important pro-oxidant in normal milk, even though it can also act as an antioxidant. As a counter effect, at least one important natural anti-oxidant, tocopherol, is concentrated in the material at the surface of the fat globules. Thus, the susceptibility of milk to oxidized flavour depends on the balance between the pro-oxidant and anti-oxidant factors at the surface of the globule.

The milk processor can take advantage of two processes that prevent or delay development of oxidized flavour: homogenization and heat treatment. The oxidative stability of milk (its ability to resist oxidized flavour) is determined to a large extent at the farm. For this reason, in recent years the main efforts to control oxidized flavour have been concentrated on the point of production.

Milk varies greatly in its susceptibility to oxidized flavour. The variability is particularly evident when the milks of individual cows are compared. Some milks develop oxidized flavour without any treatment other than refrigerated storage for as little as one day. others are so resistant that even if 5 p.p.m. of copper are added they do not develop oxidized flavour in seven days of storage. Obviously, this extreme variability is important in determining whether oxidized flavour will develop in milk and what measures must be taken to control the defect.

One of the most important factors influencing the oxidative stability of milk is the cows' feed. Pasture and other succulent feeds generally yield milk that is very resistant to oxidized flavour. The defect is encountered most frequently with dry feeds. Changes in some feeding practices in order to increase production per cow appear to be increasing the susceptibility of milk to oxidized flavour. In the United States, during seasons when the milk is susceptible to oxidized flavour, some dairymen select feeds containing high concentrations of tocopherol, or supplement rations with tocopherol to increase the oxidative stability of the milk.

The oxidative stability of milk is also influenced by some other conditions related to the cow, such as heredity, number of lactations and stage of lactation. It is helpful to recognize that such variables influence oxidized flavour, even though they cannot be manipulated to control the defect.

The cow has a marked influence on the initial oxidative stability of milk, but virtually every treatment milk receives between the cow and the consumer influences, in one way or another, the development of oxidized flavour.

Low storage temperatures favour the development of oxidized flavour. This effect is contrary to that expected of chemical reactions. Two explanations of this behaviour relate to retarding the growth of bacteria, and reducing the activity of the natural anti-oxidants in milk, but they have not been supported by experimental evidence. Storing milk at low temperatures is encouraged to maintain high bacterial quality, even though it may aggravate problems with oxidized flavour. It also frequently results in longer storage times. Thus, the adoption of mechanical refrigeration for cooling milk, particularly in bulk tanks, has sometimes been accompanied by an increased incidence of oxidized flavour in milk.

A common serious cause of oxidized flavour is contamination of milk with copper. The role of metals, especially copper, in catalysing oxidized flavour has been recognised for many years. Tinned-copper equipment has been largely replaced by 18-8 type stainless steel for the manufacture of most dairy equipment. nickel-copper alloys, commonly called 'white metal' or 'stainless metal; have been substituted for stainless steel to reduce costs, but this is an unsatisfactory compromise. White metal introduces appreciable copper contamination to milk. Copper bearing alloys must be eliminated from the cleaning circuits, as well as from milk-contact surfaces of equipment, otherwise copper will dissolve from the copper-bearing metals during circulation cleaning, and deposit on stainless steel. This copper is then picked up by the first milk that passes through the equipment after cleaning.

Another serious cause of oxidized flavour in many countries is exposure of milk to light, particularly during distribution in clear glass bottles. This may induce either oxidized flavour, or a light-activated flavour, or both. The two defects differ chemically, but in many respects the effects are similar.

Pasteurization slightly increases the susceptibility of milk to oxidized flavour. Milk pasteurized at temperature-time combinations that do not seriously impair creaming properties usually develops oxidized flavour more rapidly than non-pasteurized milk. More severe heat treatments, however, reduce susceptibility, apparently both by producing anti-oxidants and by reducing the catalytic activity of copper. Treatments that produce a cooked or heated flavour usually delay development of oxidized flavour. Products such as the non-fat pasteurized milk enriched with solids-not-fat, which is common in the United States, are usually pasteurized at temperatures appreciably above the minimum required for pasteurization (e.g. at 180-185 F for 20 sec. or longer) in order to help control oxidized flavour.

Homogenization markedly increases the resistance of milks to oxidized flavour. Many explanations have been given for this effect, but none has been supported by convincing experimental evidence. The inhibiting effect is so pronounced that homogenized milk rarely develops oxidized flavour. The popularity of homogenized milk in the United States and Canada is attributed largely to its superior flavour because of the practical control of oxidized flavour provided by homogenization.

De-aeration has been suggested to prevent development of oxidized flavour. De-aerating equipment is available that effectively removes dissolved oxygen from milk and thereby delays or prevents the defect. Unfortunately, the process is of little value commercially, because most of the bottle and carton fillers in common use allow enough oxygen to redissolve in milk to produce oxidized flavour.

Several other methods of inhibiting oxidized flavour are not useful in control programmes, either because they are not permitted by present regulations, or because they are detrimental to other quality characteristics of milk. Many anti-oxidants effectively prevent oxidized flavour when added to milk, but in most countries their use is not permitted. There are statements in the literature that bacteria growing in milk retard oxidized flavour, either by using up dissolved oxygen necessary for the oxidative reactions, or by producing anti-oxidants. Partially hydrolysed proteins have anti-oxidant properties, so proteolytic activity of bacteria could also indirectly retard oxidized flavour. However, the numbers of bacteria necessary to retard development of oxidized flavour are very large (generally over 1 million/mL.). Hence, variations in the smaller numbers of bacteria normally found in milks that meet legal standards would not be of practical importance in determining whether milk develops oxidized flavour.

Light-activated flavors

Light-activated flavour results from chemical changes in protein when milk is exposed to light. other names by which the defect has been identified include sunshine, sunlight, and solar activated. Terms used to describe the flavour are cabbage, burnt, burnt feather, burnt protein, and mushroom. Because light may induce either oxidized or light-activated flavour, or both, some investigators have not differentiated clearly between the two defects. This has resulted in confusion regarding similarities and differences between them. Several characteristics of light-activated flavour aid in understanding conditions that influence its development and methods for its control.

Homogenized milk is susceptible to light-activated flavour, but resistant to oxidized flavour. Light-activated flavour increases with intensity and duration of exposure, and the predominant flavour characteristic changes during exposure and subsequent storage. The temperature of the milk during exposure and storage is also critical. Higher exposure temperatures (50-60 F) result in more pronounced flavour, but the flavour of samples exposed at low temperatures (35-40 F) may increase in intensity for several hours after exposure. If, after exposure, the milk is held at higher-than-normal refrigeration temperatures (e.g. 50-60 F), the flavour may decrease in intensity. The time interval between homogenization and exposure also has an influence. If milk is stored one or more days after homogenization, but before exposure, it is much less susceptible to the defect.

Light of the shorter wavelengths (blue and green) has greater effect than that of longer wavelengths (yellow and red). Fluorescent light is more troublesome than incandescent light, because it usually has a higher proportion of its energy in light of short wavelengths. Also, because fluorescent lamps are not hot, milk is frequently placed close to them in retail display cases.

Selecting fluorescent lamps that emit a high proportion of the 'warm' tones helps to minimize damage from fluorescent light in display cases. Likewise, use of colored bottles or paper containers that absorb much of the light of shorter wavelengths provides protection, but does not necessarily prevent activated flavour under very adverse conditions.

The cows' feed influences light activated flavour as it does oxidized flavour. Green feeds appear to yield milks with greatest resistance to the defect. Hence, as for oxidized flavour, the susceptibility of milk to light activated flavour varies seasonal]y and is greatest during the winter.

Rancid Flavour

The term 'rancid', when applied to milk and other dairy foods, refers to a flavour defect caused by hydrolysis of fat, rather than by oxidation of fat. Milk always contains an enzyme (or a group of enzymes) known as lipase, which is able under certain conditions to hydrolyse fat, splitting off fatty acids that are responsible for the rancid flavour. Freshly drawn milk from healthy cows is never rancid. Depending on conditions, rancidity may develop on aging of raw milk. Pasteurization destroys lipase, so properly pasteurized milk will not go rancid.

In most milks, the so-called 'membrane' around the fat globules appears to protect the fat from attack by lipase. Certain treatments, known as activating treatments, change the fat globule surface sufficiently to permit the lipase to act on the milk lipids and produce rancid flavour. Three activating treatments that may be encountered in milk production and processing are

1. Agitation of warm milk, particularly under conditions that produce foam;
2. Homogenization of raw milk (or mixing raw and homogenised milk); and
3. Temperature fluctuations such as cooling milk, warming it to about 86 F and then cooling it again.

When milk samples from individual cows are cooled and stored one or two days, a small proportion of the samples may develop rancid flavour. This rancidity is sometimes referred to as spontaneous, because no treatment other than cooling is required for its development. Some investigators believe spontaneous rancidity is caused by a lipase which differs from that present in normal milk, whereas others consider that the fat globules of 'spontaneous' milk do not have sufficient protective matter on their surfaces to prevent attack by lipase. Regardless of the explanation, spontaneous rancidity is not a serious source of off-flavour in commercial milk supplies. It occurs most commonly with milk from low-producing cows that are very advanced in lactation and should be dried off. Also, if milk from such cows is mixed with non- spontaneous milk, the mixture usually does not develop rancid flavour unless it receives an activating treatment. Hence, when rancidity is encountered in commercial milk, the immediate cause is usually an activating treatment.

Susceptibility of milk to induced rancidity varies greatly. Rancid flavour of differing intensity develops in milk samples taken from individual cows in a herd, and even from individual quarters of a cow's udder. Individual cows show marked daily variations in susceptibility of their milk. Susceptibility is often greater when cows are in advanced lactation. Green succulent feeds make milk more resistant than dry feeds. The numerous variables that influence susceptibility are at least partly responsible for sporadic and seasonal variations in incidence of rancidity.

To achieve labour efficiencies, the producer adopted the bulk collection system and changed from bucket milkers to pipeline milking systems. The average number of cows on dairy farms increased, and to increase milk yield per cow farmers depended more and more on feeding hay and concentrate mixtures.These progressive developments seriously aggravated rancid flavour problems. The change is not attributable to a single cause, but to a combination of changes that took place simultaneously.

When the bulk tank system was adopted, it was possible to cool milk more rapidly and store it at lower temperatures than with the can system. Rancid flavour induced by some activating treatments develops more rapidly as the storage temperature is decreased. In some instances, defective or inadequate refrigeration systems resulted in prolonged agitation of warm milk, or temperature fluctuations resulting from additions of warm milk to previously cooled milk. In some districts, milk was collected only on alternate days, resulting in longer storage which favored the development of rancidity.

When new installations were made, most were entirely of stainless steel in the milk-contact areas. This eliminated sources of contaminating metals. The lipase in milk is unstable in conditions that favour oxidation. Thus, eliminating copper contamination, added to the rapid cooling of milk, increased the stability of the lipase.

Many early installations of pipeline milkers provided conditions which subjected milk to activating treatments. Milk was commonly elevated in vertical sections of the vacuum line by bubbling air through it. Excessive amounts of air entered the systems at milker claws, fittings, etc. Pumps were operated continuously, but starved. Low producing cows in advanced lactation were not so readily detected and were retained in the milking strings too long.

As the number of cows per herd was increased, farmers could no longer supply enough pasture or silage for the cows. Hay and grain concentrate mixtures provided increasing proportions of the nutrients for milk production, and these dry feeds yielded milk with greater susceptibility to rancidity. All of the above conditions probably contributed to rancid flavour problems. To control the defect, many unsatisfactory practices and equipment installations have been eliminated. Although most commercial milk is undoubtedly more susceptible to rancidity than that produced in pre-war years, the industry has learned to cope with the problem by minimizing activating conditions.

The pasteurization, by inactivating the lipase, provides the processor with a very effective method of controlling rancidity. High susceptibility of milk to rancidity may limit flexibility of operations in a processing plant by necessitating prompt pasteurization to prevent development of rancidity. In properly pasteurized milk, rancid flavour should not be a problem. Processing plant employees must be aware of conditions that may induce rancidity. If cooled milk is warmed for separation, the cream should not be cooled again without being pasteurized. Homogenized milk must be pasteurized before, or immediately after homogenization Homogenized products should not be mixed with raw products. Prolonged agitation of warm raw milk (80-120 F) or pumping it through starved pumps may induce rancidity.

ACID DEGREE VALUE (ADV)

Rancidity is characterized by the release of free fatty acids because of the action of the lipase enzyme. When consumers taste rancid milk, they are detecting the short chain fatty acids. An Acid Degree Value test can be used to measure the presence of long chain fatty acids. There is about a 70 percent correlation between flavor and ADV. ADVs are more meaningful for raw milk samples from individual farms. In mixed milk on trucks or in silo tanks the effect of milk from one farm is minimized. Rancid milk from one farm may cause a rancid flavor in the entire tank. Therefore, ADV results from mixed milk are not very meaningful.

Milk samples from individual farms should have ADVs of less than 0.80. When results exceed 1.00, the milk will have a soapy-bitter taste. Some farms may have ADVs above 1.00 at the time of processing. ADVs should be conducted at the time the milk is processed rather than when the milk is received at the plant. When ADV results exceed 1.20, one can expect consumer complaints and a negative effect on consumption and sales. Action should be taken when samples results are above 1.00.

FARM OXIDIZED FLAVOR

Oxidized flavor is also a reaction of milkfat. Milk with a cardboardy or metallic taste is more common in milk during the winter and early spring. The off- flavor can be detected in raw milk, but sometimes not until two days after collection. It can also be a problem in any pasteurized milk or dairy product that has not been flavored. Causes are different than for the light induced flavor of milk purchased at stores, although the taste is similar.

Increased susceptibility of milk to the chemical development of oxidation is due primarily to less antioxidant in the milk. The main cause is the decreased amount of vitamin E, an antioxidant, in stored forages, which reduces the amount found in milk.

Once milkfat has begun to oxidize, the intensity will continue to increase overtime. The taste may not be apparent in the milk, but may be detected in high fat products such as butter or vanilla ice cream. In the majority of cases, oxidized flavor problems start at the farm.

There are several causes of oxidized flavor on farms or in processing plants. They include:

- Equipment surfaces improperly cleaned.
- Minerals in the water supply.
- Acid water and copper tubing.
- Use of chlorine sanitizers.
- Stored forages low in vitamin E.
- Feeding high levels of vegetable fats, i.e. soybeans, cottonseeds.

It is usually necessary to correct all causes of the problem before the off-flavor will disappear. Feeding extra vitamin E will not overcome excessive copper or iron in the water supply used to wash equipment or compensate for dirty milking equipment. There are corrective and preventative measures that can be taken. They include the following:

- Wash all surfaces that are exposed to milk after every use. Any milkfat or protein left in a bulk or storage tank, pipeline, milker unit, or pump oxidizes readily. Check all equipment on the farm, in truck tanks, and receiving rooms to be sure that they are clean.

- The presence of copper or iron in the water used to wash equipment will cause oxidized flavor. As little as 0.1 ppm copper or iron may cause a problem. Unless another water supply is available, treatment is necessary.
- Water in mining areas may be acidic. When copper tubing is used to transport the water to the milk house, some copper may be removed. It will settle on equipment surfaces and cause oxidation when milk comes into contact. Copper tubing should not be used for water in the milk house if the farm has acid water.
- Sanitizers are oxidizing agents, so solutions should be completely drained from all equipment. Chlorine precipitates metals, so iodine sanitizers are recommended for equipment when an oxidized flavor problem is present. However, it is not appropriate to rinse with water after sanitizing.
- Provide some green feed, i.e. pasture or greenchop to milking cows from May to October. These contain high levels of beta- carotene and vitamin E.
- Check the ration for adequate protein content. It should be appropriate for the level of milk being produced. Check that the milk protein content is normal.
- Ingredients that increase the proportion of unsaturated fatty acids in the milk may increase the milk's susceptibility to oxidation.
- Nutritional problems, which depress the milkfat percent, often increase the unsaturation of the milkfat and make it more susceptible to oxidation. To correct milkfat depression, evaluate the ration for fiber levels, particle size, fat content and source, and the cereal grain source (i.e. heat-treated, particle size, high moisture). Milkfat test is considered depressed when it falls more than 0.3 percent below breed average.
- Light exposure of milk in a glass or plastic milk line will not cause an off-flavor problem. The exposure time to sunlight, daylight, or fluorescent light is too short.
- When obvious causes have been corrected and the problem persists, add 1,000 or up to 7,000 International Units of vitamin E per animal daily for one to two weeks. This should be done only in extreme cases. Consult with a nutritionist before feeding these levels.

Feed Flavor

All feed flavors are absorbed through the cows system rather directly into the milk. Cows impart an odor and taste within 30 minutes of eating or breathing silage. It is strongest after about one hour. The odor and taste of grass or corn silage, legume hay, and brewer's grains are the ingredients causing most of the problems.

The two methods which odor can be transferred to milk is:

- Nose or mouth ! lungs ! blood ! milk
- Mouth ! digestive tract ! blood ! milk

Unclean Flavor

Some farm milk samples have an unpleasant, dirty after taste. This problem often occurs during winter. Frequently it is an absorbed flavor, like silage. Usually cows breathe air with a barny odor and transfer it to the milk. Dust, dirt and manure can cause an unclean flavor of milk. Cows and their surroundings must be kept clean. Milking equipment that has not been properly cleaned and sanitized may be a factor.

Washing a cow's udder with water and failing to dry them is one of the primary causes of unclean flavors. Be sure excess moisture is removed. In a small to average sized herd, a few animals with ketosis will cause an unclean flavor. This odor and taste is objectionable.

Follow these suggestions to prevent unclean flavor problems in milk:

- Keep floors, walls, and ceilings of milking and housing areas reasonably clean.
- Provide adequate ventilation to eliminate stale odors.
- Clip hair from the udder, teats, and flanks of milk cows.
- Provide adequately bedded stalls so that cows may lie down.
- Wash udder and teats with a sanitizer solution and dry prior to attaching milker units.
- Clean and sanitize all milk handling equipment between uses.
- In some circumstances it may be necessary to withhold milk from ketotic cows.

Malty

Milk is an excellent growth medium for bacteria. It provides the nutrients and moisture and has a near neutral pH. Off-flavors are the results of bacterial growth (psychrotrophs). The type of bacteria is more important than the number. Psychrotrophs include many kinds of bacteria, all of which cause spoilage. Their ideal growth temperature is 65-70°F. Cold temperatures slow their growth, but do not kill them. At 45°F or above, bacterial growth is steady and off-flavors may be present in two to three days. Malty flavor tastes like Grapenut cereal. The cause is due to Streptococcus lactis in poorly cooled milk. The problem can be exacerbated when the milk from one bulk tank is mixed with that from many others in a truck tank, then in a plant raw milk storage tank. A sample of milk from a farm bulk tank not cooled for 12 hours will always have a high bacteria count. Malty flavor is generally a forerunner of a high acid flavor. It rarely develops in pasteurized milk. However, the characteristic flavor will remain after processing, although the flavor developed in raw milk. If not stopped by pasteurization, a malty flavor will later become high acid.

High Acid

Bacterial flavors are not acceptable in milk for fluid or manufactured purposes. Sanitation is the key to preventing high acid flavors. Spoilage is due to bacterial action on lactose (milk sugar). All milk handling equipment surfaces should be washed after each use and sanitized just prior to reuse. This includes washing all milking units, pails, and pipelines on a twice per day basis. Bulk tanks must be washed each time they are emptied.

Milk from the bulk tank should be picked up at least every other day. Longer time periods between pickup and partial emptying of a tank causes problems. Any milk left in a bulk tank for more than three days has the potential of high bacteria counts and off-flavors in fluid milk and manufactured products. Depending on milk composition, normal acidity test results can vary from 0.13 to 0.17 percent lactic acid. Before a hint of acid taste can be detected, this will usually rise to above 0.20 percent. Milk can safely be used for fluid purposes if acidity does not exceed this and there is no acid taste at the time of pasteurization.

Putrid

Psychrotrophs cause flavors that are often described as stale, lack freshness, fruity, bitter, fermented or putrid. Frequently the titratable acidity may be near normal. Putrid flavors are the result of bacterial contamination, storage temperature above 40°F, and age. Spoilage of the milk is by bacterial action on the protein rather than on the lactose. Putrid milk will curdle, separate, and may smell rotten if left for a few days.

The most common source of the problem is contamination in processing plants following pasteurization. Psychrotrophs do not generally survive pasteurization unless large numbers of the bacteria are present in the raw milk. Usually the contamination occurs because some part of the pumping, holding, and filling system is not properly cleaned and sanitized. Putrid flavors may originate in raw milk caused by contamination and holding raw milk for three or four days after collection from farms. Occasionally, retail samples of milk are found with putrid flavors and excellent bacteria counts. This can result when bacteria counts in the millions are present in raw milk at the time of pasteurization. Proper pasteurization and packaging without contamination eliminates most of the psychrotrophs. Although all the bacteria are destroyed, the spoiled flavor remains. No heating, vacuum treatment or other processing procedure will lessen the flavor.

HEATED FLAVORS

In commercial processing, milk is almost invariably subjected to heat treatments at least as severe as the minimum required for pasteurization, and frequently more severe. In general, the intensity of the heated flavour increases with that of the heat treatment.

Pasteurization produces a slight heated or cooked flavour in milk. This process is used so widely that most consumers consider a slight heated flavour to be typical of normal milk and they do not find it objectionable. A slight heated flavour contributes to the apparent sweetness and richness of milk masks some off-flavors of low intensity, and may make milk less susceptible to oxidized flavour. For these reasons, some processors deliberately use pasteurization temperatures appreciably in excess of the legal minimum to produce a slight heated flavour in the freshly pasteurized milk. This flavour decreases in intensity during storage. Hence, although uniformity is desirable, the intensity of the heated flavour does not remain constant during the time milk is in distribution channels.

Heat treatments at much higher temperatures and for longer times than required for pasteurization are used in the processing of many dairy products, and they produce pronounced heated flavors with differing predominant characteristics. The first that becomes evident results from liberation from the proteins of hydrogen sulfide and other sulfides. More severe heat treatments give scorched and burnt flavors caused by more extensive changes in the proteins; caramel flavors attributable to reactions involving lactose; and lactone or; coconut-like flavors resulting from changes in the milk fats. These more severe heat treatments are accompanied by changes in the tactile. or mouth-feel, characteristics of milk. Such flavors are typical of sterilized milks, and are objectionable to most consumers. Hence, extensive research has been directed towards sterilizing milk without producing the objectionable flavour caused by conventional sterilizing treatments.

Bacterial Flavors

Off-flavors caused by the growth of bacteria in milk are not detectable until large numbers of bacteria are present, usually millions per milliliter. Hence, the milk would not meet legal standards for bacterial quality. Nevertheless, defects caused by bacteria are encountered from time to time, and it is important to know their characteristics and conditions under which they develop. Milk is such a good food for bacteria, as it is for man, that it is very subject to spoilage. It must be rigorously protected from bacterial contamination, and kept cold to minimize growth of bacteria that are present. If flavors of bacterial origin develop in raw milk, this indicates that sanitary practices have been inadequate, or that the milk has been held at too high a temperature, or too long. A sour, high acid, or malty flavour may result if raw milk is held at temperatures above 45 F. At lower temperatures, the flavors that develop are usually caused by psychrophilic bacteria and are described as fruity, bitter, rancid or putrid. Pasteurization destroys the milk souring bacteria so effectively that, when pasteurized milk spoils from microbial growth, the cause is usually bacteria that contaminate the milk after pasteurization. Thus, the keeping quality of pasteurized milk stored at 45 F or below can serve as an indication of sanitary practices in the plant. As for raw milk, the psychrophilic organisms that grow at these low temperatures cause fruity, bitter. rancid or putrid flavors. Properly pasteurized milk, adequately protected from post- pasteurization contamination and stored at 45 F, should resist psychrophilic spoilage for at least two weeks.

2

Cause of Taste Defects in Milk

DEFECTS DUE TO PRESENCE OF ABNORMAL MILKS

These are colostral milks or pathological milks which normally should not be collected or marketed. These milks have an abnormal mineral and protein composition (high chloride content - low casein content) which gives them a characteristic salty taste. Their presence in low concentrations in the collected milk does not change the organoleptic characteristics of the mixture and therefore cannot be detected by sensory analysis. The adulteration can only be detected by specific instrument methods which are chemically or physically based.

DEFECTS DUE TO CHANGES IN MILK CONSTITUENTS

The main alterations in milk taste are due to bacteriological, chemical and physical changes in the milk constituents. These three aspects will be studied in this chapter.

Defects of Microbial Origin

Milk is a highly favourable culture medium for microbial life because of its chemical composition, which is rich in various nutrients, and because of its water content. Thus it is highly vulnerable to any contamination that can occur in production, processing and marketing.

Serious taste and odour defects can appear due to an accumulation of products resulting either from cell metabolism or from the effect of complex enzyme systems on the milk constituents. Many undesirable changes in organoleptic quality are possible when environmental conditions are conducive to microbial proliferation and enzyme activity. The following list of adverse changes is not restrictive. Most frequently one speaks of milk that is sour, bitter, fruity, rancid, malty, with an off-flavour tast, and also of dirty milk, etc. These forms of spoilage are associated with the growth of yeasts, moulds and bacteria. In view of its ecological characteristics, bacterial contamination is the most frequent and the greatest, and its potential development should be feared most of all. This contamination is responsible for two main types of defects: souring and lipolysis.

The defects due to acidification are the most frequently encountered since, lactic flora is one of the main natural contaminants of milk, being predominantly mesophilic in character. Presence of this flora is demonstrated by a greater or lesser rapidity of acidification depending on temperature conditions at the production, processing and distribution stages. This form of spoilage is apparent in the sour taste, due not to lactic acid (which is barely volatile because of its low vapour pressure), but to the presence in smaller quantities of organic acids (acetic acid, propionic acid, carbon dioxide).

Furthermore, the heterofermentative character of lactic flora is demonstrated by a more or less marked release into the environment of numerous different volatile compounds (aldehydes, ketones and alcohols) besides the organic acids. These compounds, when they develop in limited quantities, are sometimes sought after since, they help to form the typical aroma of many milk products. On the other hand, when they are present in high concentrations they produce an unpleasant taste and odour. A classic example is diacetyl, which in a highly diluted state is responsible for a nutty taste and in a more concentrated state results in decided bitterness. Likewise, different tastes designated as malty, grape or caramel, have been attributed to the proliferation of Streptococcus lactis maltigenes.

Lactic bacteria generally are not heat resistant and most of them are destroyed by low temperature pasteurization. However, the survivors or recontaminating bacteria can be responsible for further souring if temperature conditions are favourable to their development. The predominant lactic flora in milk is mesophilic and the natural environmental conditions in warm and hot regions are often favourable to its proliferation. The use of cold treatment at different stages of production, processing and marketing makes it possible in practice to decrease considerably the dangers of spoilage due to uncontrolled proliferation of these lactic bacteria. However, it should be noted that in these conditions even with cold techniques the multiplication of microorganisms and enzyme changes are not wholly blocked and that irreversible changes in the organoleptic quality of the product can still occur.

The defects due to uncontrolled proteolysis have their origin in bacterial flora which contaminates milk on the farm and can induce proteolysis of varying intensity depending on the species concerned, time periods and temperatures. In practice, mesophiles in the case of chilled milk and psychrotrophic bacteria in the case of refrigerated milk can be the cause of milk protein solubility of varying intensity because of their extra-cellular proteases. The protein fractions released (peptides, aminoacids and amino-ammonia) can be responsible for the development of abnormal bitter, ammonia or putrid tastes. Moreover, although most proteases are heat-sensitive, some enzymes coming from psychrotrophic bacteria are thermostable and their harmful effects can appear even in standard sterilized products. Therefore milk products, although sterile, can spoil during storage due to residual enzymes that have not been rendered inactive.

Many findings have shown that susceptibility to lipolysis is not the same for all milks. Two types of factors determine the frequency and intensity of defects, viz. enzyme activity as such and the condition of the fat substratum.

The enzymes responsible for lipolysis have two different origins. Fresh cow's milk contains several lipolytic enzymes, membrane lipase and plasma lipase. Although their isolation and characteristics have not yet been totally elucidated, it is generally recognised that these enzymes have little effect on raw milk when milk is in the resting stage and the fat globules are intact. In milk products which have been subjected to heat treatment for pasteurization (72°C for 15 seconds) or sterilization (classic or UHT), their activity is believed to be nil because they have been rendered inactive by heat treatment. However, recent studies have revealed cases of enzyme reactivation causing defects.

The second origin of lipases is associated with unsatisfactory hygienic conditions of the milk on the farm. Very many bacteria species can produce lipases. Psychrotrophic bacteria are the greatest threat because of their ability to develop at low temperature and their high extra-cellular activity.

The most active and most frequently encountered microorganisms and bacteria species have a relatively short multiplication period (4 to 24 h) at commonly applied temperatures (0 – 49C); when the temperature rises the growth rate is multiplied by 5.8 for every 10°C rise.

Correlations between bacterial density and the appearance of defects have been established. In milk, when germ concentration exceeds 10^6/ml off-tastes can be detected. When the fat content is partly concentrated (cream, butter, cheese and powders) defects can emerge at lower thresholds, especially during prolonged storage.

Many extra-cellular lipases of psychrotrophic bacteria are also characterised by extremely high heat resistance; they are only partly inactivated by the heat treatment used in UHT pasteurization and sterilization, although the bacteria themselves may be destroyed. Organoleptic defects may therefore appear even in properly heat-treated microfree media.

Frequency and intensity of lipolysis increase when the initial globular state of the fat is modified. The influence of endogenous factors traceable to the cow (effect of feeding - effect of lactation stage) which determine the synthesis of the globular membrane and the stage of lactation has been demonstrated. In most cases, however, factors extraneous to the animal have a dominant role in the development of lipolysis. Excessive shaking, addition of air, repeated thermal shocks, and homogenization, all of which can occur at different stages of production and processing adversely affect the integrity of the fat globule, modify the interfaces between the fat and non-fat phase and lead to an increase of lipolysis. In the light of many recent studies it appears that the development of milk production systems - characterised by the expansion of collection areas, increased mechanization of milking, collection and transport - leads to longer processing times. These factors have overall led to a more marked degree of spoilage of the raw material of milk than in the past and to an increase in organoleptic defects in the finished product.

To benefit fully from cold treatment, this must only be applied in environments where few psychrotrophic bacteria are present. Quality improvement of milk at the production stage, where the aim is to inhibit contaminant germs by observing strict rules of hygiene, is therefore inseparable from the rational utilisation of cold treatment.

Defects Due to Fat Oxidation

Oxidized flavour is a result of the effect of oxygen on the milk phospholipids which causes the development of various flavours - cardboard, metallic, fish or oily. Oxidation of polyunsaturated fatty acids of the phosphatides of the globular membrane induces the formation of hydrogen-peroxide. The results of the reaction are the formation of alcohols, acids, aldehydes and ketones producing the above-mentioned defects.

The presence of contaminant metal traces (iron and copper) and also energy radiation exposure (light or radioactive) catalyses oxidative reaction. In practice, in order to reduce these oxidative defects it is advisable to decrease metal contamination by using appropriate materials made of stainless steel or glass and handling the milk in recipients providing effective protection against sunlight or artificial light. The replacement of trapped air in the product and in the space at the top of the container by an inert atmosphere ($CO_2$and nitrogen) or by a vacuum is also highly recommended in the case of food products with a high fat content which must be stored for lengthy periods at ambient temperatures.

In the specific case of recombination of milk products the use of iron- and coppercontaminated water should be forbidden and the addition of air during mixing and transfer operations should be prevented. Low pasteurization can intensify the emergence of defects, but stronger heat treatment has the opposite effect; it causes the release of sulfydryl (-SH) groups originating from lactoserum proteins which have an anti-oxidant effect and prevent the formation of off-taste. Chemical antioxidants have the same positive effect; however, their addition is generally forbidden by law. In some countries ascorbic acid can be added to supplement the milk which in low concentration and in the presence of copper stimulates oxidation, but at higher concentration prevents the development of oxide flavours.

Homogenization provides effective protection against oxidation. Significant changes in the globule surface and in the composition and structure of membrane proteins are probably responsible for this improvement in flavour stability. In the specific case of milk products made with recombined milk, it is essential to obtain a stable emulsion of the fatty matter in order to prevent rapid spoilage. The observance of optimum homogenization conditions is of primary importance in this case.

It should also be noted that the defects resulting from oxidation of fatty acids can develop in non-fat products when they contain trace phospho-lipids. In all cases low-temperature storage of products containing milk fat provides an additional guarantee against oxidation spoilage.

Defects Due to Heat Treatment

Most milk constituents can be physically and chemically altered by heat treatment used for preservation. The extent of the resulting changes will increase with the duration and temperature of heat treatment, but it also depends on the specific sensitivity (which can vary) to heat of the respective milk components; these can be listed, in decreasing order of importance, as free enzymes, serum lactoproteins, linked enzymes, phospho- and caseinocalcium complexes, lactose and lipids.

Heat treatment completely inactivates enzymes and destroys the most heat-resistant microorganisms but causes major changes in product characteristics, which generally decrease acceptability. For example, alteration will be slight for heat-treated and pasteurized products, slightly intensified for UHT sterilized products, and more marked for products that have been subjected to the conventional lengthy autoclave sterilization.

As a rule inactivation of most enzymes is obtained through normal heat treatment and pasteurization (60–100°C) which give a better stability to heat-treated products, and eliminate in particular organoleptic defects due to the development of uncontrolled proteolysis and lipolysis in UHT-treated long-life products. However, the inactivation of particularly heat-resistant enzymes may not be complete. This is the case, for example, of lipases and of thermolabile phosphatases when protected by the fat phase and/or the protein phase during heating which are later reactivated during storage. Freak organoleptic properties due to this type of lipolysis and proteolysis have been reported quite frequently, especially for milk products derived from milks that are initially highly contaminated with psychrotrophic bacteria.

Many changes in heated milk characteristics are the result of direct spoilage of lactoserum proteins or interactions involving soluble milk proteins following heat treatment.

Direct modifications are the development of sulfhydryl (-SH) and hydrogen sulfide (H_2S) groups from â - lactoglobulin, and also the formation of small quantities of free sulfides and mercaptans. These impart a burned taste, but it should be noted that these free groups also decrease the oxidation-reduction potential and give a better protection of lipids against oxidation.

Recombined milk products frequently exhibit this burned taste, which is usually considered unpleasant. Its intensity increases with the heat-induced denaturation that the milk undergoes during its transformation into powder, and also with the additional heat treatment to preserve it following recombination. The development of this taste is therefore in direct ratio with the solubility value of the powder selected and the time-temperature relation of the heat treatment applied to the recombined product to ensure its stability for the desired length of time.

Thus the taste of a pasteurized product prepared from a "low temperature" powder will be little changed. On the other hand, the change will be more marked for a milk sterilized by the autoclave process and obtained through recombination with a "high temperature" powder.

The main interaction between lactoserum proteins and casein is the formation of a complex between the â -lactoglobulin and Kappa-casein. This occurs when more than 50 percent of the lactoserum proteins are denatured. It has many consequences, but where organoleptic quality is concerned, it retards the release of the R-SH and H_2S groups when additional heat treatment is applied. Heating changes the salt balance towards insoluble forms. Changes in the composition of micella surfaces and phosphate precipitation decrease the stability of the colloidal phase but these consequences have little influence on organoleptic quality.

Lactose is generally not affected significantly by UHT pasteurization and sterilization, but the damage is greater when more drastic heat treatment is used. In this case a decomposition of the lactose occurs and acids form, in particular formic (50 to 75 percent of the newly formed acids), lactic, acetic, pyruvic, propionic and butyric acids; hydroxymethylfurfural and furfuraldehyde also develop. All these components help in forming the taste, odour and colour of milk but their respective roles have not been clearly determined.

The main change occurring from heating of milk products còntaining lactose corresponds to a non-enzymatic browning reaction or Maillard reaction. This reaction, which requires little activation energy, is self-catalysing; in its initial phase condensation occurs between a free aminic group of the casein, especially the lysine group and the lactosereducing aldehyde functions. It leads to the formation of a Schiff base followed by the Amadori rearrangement and degradation into a brown pigment (melanoidins). The browning does not always appear during heat treatment since, the Schiff bases so formed can become degraded slowly in dry, liquid or medium-moist milk products during storage. A caramel-like taste develops.

In practical terms, in order to minimize the browning reaction the amount of heat treatment must be limited and storage times and temperatures reduced. The use of certain additives like sodiumbisulfite, sulfur dioxide, and formaldehyde, or the presence of (-SH) groups in the environment can effectively help to prevent these defects. Under the influence of intense heating the fat produces lactone and methylketone compounds which if highly concentrated can cause undesirable flavours (coconut flavour) in concentrated and dried milk. but at lower levels they are responsible for the characteristic and much sought-after flavour of food cooked in butter. In moderately heat-treated liquid milk products their contribution to the development of freak organoleptic effects appears to be less.

DEFECTS THEIR CAUSES AND PREVENTION

Condensed milk and evaporated milk are the products suitable for prolonged storage.They have typical properties after production by which they are identified. These properties must be such that the product is fit for sale immediately after production and does not alter during reasonable period of storage. They should, therefore be physically, chemically and bacteriologically fit for human consumption by the end of storage period. Routine examination of the product soon after manufacture as well as during storage may be carried out to judge the quality of the product. If any defect is noticed, proper care should be taken to climate the defect in the subsequent batches. It is, therefore, important to know what type of defects may occur in the product, the reasons for their occurrence and the preventive measures to be used to avoid these defects.The defects which may occur in condensed milk may be divided into two categories as follows:

MICROBIAL DEFECTS

These defects are due to presence of abnormal bacteria in the milk or in products.

The common defects are as follows:

- *Gassy fermentation/Bloats:* Formation of gas takes place in cans and barrels of condensed milk causing bulging or bursting of containers. Gas producing yeasts are the cause of many types of gaseous fermentation. The source of contamination may be the raw milk, or inferior quality sugar or unhygienic factory conditions especially not properly washed and sanitized equipments and filling machine. To avoid this defect good quality raw milk and properly preheating temperature should be used. Only good quality sugar without any yeast contamination should be selected. Proper sanitary conditions should be maintained during manufacture and packaging of condensed milk. The containers should be filled fully with little space for air or oxygen.
- *Bacterial thickening:* Condensed milk gets thickened progressively during storage. This is due to microorganisms which produce rennin like enzyme.These organisms are easily destroyed during preheating

process. Optimum sugar ratio (64.5) may inhibit the growth of microorganism. Low temperature storage also helps in reducing the bacterial thickening.

- *Mould buttons:* This defect occurs during storage due to mould contamination. Small reddish brown pieces of curd about ¼" to ¾" in diameter are formed on the surface causing localized coagulation. This defect occurs after storing the product for some time. The causative mould "Aspergillus repeno" produces a"clotting enzyme" which causes localized clotting. High temperature storage also helps in the growth of the organism. The milk may be infected with the organism during concentration process.

To avoid this defect, scrupulous cleaning and care of dairy equipments is essential.Since this organism does not grow at low temperature storing of condensed milk at lower temperature will prevent this defect.

NON MICROBIAL DEFECTS

The non-microbial defects are of chemical or physical origin. These are listed below.

- *Sandiness:* Good quality condensed milk should possess a smooth homogenous texture and be pleasant to palate. Some times however the milk may be gritty containing large number large sized hard lactose. The solid particles are of such size that the product lacks smoothness and grittiness is noticeable, as the sample is being tasted. This defect is readily detected by an average consumer.Sandy, rough grainy, granular, and gritty are the term used to describe this defect.

The sandiness may be due to presence of relatively large size crystals of lactose.Also if excess amount of sugar is used in manufacture of condensed milk, sugar particles may also crystallize out and cause sandiness. If manufacturing conditions are not conducive to the formation of small lactose crystals, large and coarse crystals are formed. Cooling of condensed milk must be carried out in such a manner that smooth texture is obtained by the formation of large number of minute sugar crystals.High viscosity also delays the crystal formation. It is, therefore, necessary to rapidly cool the condensed milk in the initial stages.Correct cooling and induced rapid crystallization with correct amount of seed lactose will help to avoid this defect. Optimum storage temperature is also essential to avoid this defect.

- *Age thickening:* Thickened condensed milk is the most common defects seen in the sweetened condensed milk. This defect varies markedly in its intensity from slight jelly to a firm and consistency. The defect becomes progressively more intensive upon storage, especially at room temperature or above.Preheating temperature of milk and degree to which the milk is concentrated have been observed to have profound effect on age thickening. With high prepackaging,heating temperatures, there is a greater tendency upon the product to thicken early. With increasing concentration of milk solids the thickening tendency becomes more marked. In order to avoid early thickening of condensed milk,optimum pre heating temperature should be maintained. Sugar should only be added at the end of condensing of milk. The product should preferably be stored at temperature below 15°C with the addition of proper type of stabilizers age thickening may be decreased to a great extent.

- *Brown colour:* usually brown colour discolouration is associated with age thickening, both of which become progressively more intense on storage. The acidity and temperature of storage are factors, which determine the rapidity of change. This defect may be avoided if the condensed milk cans are stored at reasonably low temperatures. Other defects in condensed milk include some flavour defects such as rancid, tallowy, metallic etc. Following good manufacturing practices, good quality raw material, good quality milk, good packaging and storing the product at low temperature may help to avoid the defect.
- *Fat separation in condensed milk is rare. Defects in Evaporated Milk:* Evaporated milk is sterilized product. If sterilization is properly carried out the product will remain in good condition without any bacterial spoilage during storage. However, if sterilization process is improperly carried out, some microorganisms and spore may, survive and cause spoilage, during storage, Gassy fermentation, coagulation, bitterness and fishy flavour are some of the defects noticed in improperly sterilized evaporated milk. If the bacterial defect in evaporated milk is to be eliminated, it is essential that the sterilization process should be carried properly and adequately with strict cleanliness in the factory.

Chemical/Physical Defects in Evaporated Milk

- *Age thickening:* As in the case of sweetened condensed milk thickening or high viscosity of the product is due to improper preheating of milk, higher total solids concentration, inadequate homogenization and storage at higher temperature. To eliminate this defect the milk should be heated at optimum preheating temperatures. The concentrated milk should be homogenized at proper pressure and the product should be stored at low temperature.
- *Age thinning/low viscosity:* Milk like consistency of evaporated milk indicates its low viscosity. This defect may be due to insufficient concentration of milk and inadequate or no homogenization of milk or improper sterilization. This defect is eliminated by following

correct/adequate processing conditions and storing the product at low temperature.

- *Fat Separation:* A thick, heavy cream layer is noticed at the top of the can when it is opened. The cream layer is very dense and will not easily mix with remainder of milk. Even after mixing creamy chunks of butter particles are noticed floating in the milk of relatively low viscosity. The reasons for this defect to appear are in adequate homogenization high storage temperature, long storage period and improper handling during storage. Proper homogenization, low temperature storage and correct handling of the product will eliminate this defect. A more viscous is desirable from fat separation point of view.
- *Sediments/Mineral deposit:* Sediments noticed in evaporated milk cans may be due to crystallization of some of the calcium and magnesium salts, and denatured proteins. This gritty sediment formation takes place during prolonged storage of evaporated milk. The rapidity with which the sediments are formed is influenced by the nature of the milk, conditions of manufacture and temperature of storage. Selection of good quality milk, correct processing conditions and low temperature storage will help to eliminate this defect. Higher viscosity product is less prone for this defect.
- *Brown Colour:* The brown colour in evaporated milk is associated with high sterilization temperature, high storage temperature and longer storage period.The sterilization temperature should be as low as possible with germicidal efficiency. The brown colour defect in evaporated milk is eliminated by employing proper preheating temperature, correct sterilization process and low temperature for storage of the product.

DEFECTS DUE TO TRANSMITTED FLAVOURS

Because of its high water and liquid content, milk can be a major vector of fat-soluble and water-soluble foreign substances which cause off-tastes. These substances can come from the feed given to the cow, as well as from the environment before and after the milking process. They can be transmitted either indirectly in the full udder through the respiratory and/or digestive system, or directly by contact with the product after milking.

Transmission through the respiratory system is due mainly to volatile compounds contained in the atmosphere of badly ventilated premises (gases from feed or manure; gases eructed by healthy or diseased animals and the presence of foreign chemical substances, e.g. fossil fuels, disinfectants). It has thus been possible to prove the origin of various odours or tastes of cabbage, ensilage, cow, and farm. When the source of the flavour disappears and the animals once again breathe wholesome air the foreign substances accumulated in the udder are rediffused in the bloodstream and are expelled via the lungs.

The feed consumed by the animal is another known cause of unpleasant tastes and odours. When the cow consumes fresh or preserved strongly flavoured forage (ensilage, cabbage, wild tuber plants, etc.) two to four hours before milking, their characteristic taste and odour appear in the milk and can persist until elimination or metabolisation of the feed, or sometimes up to about 12 hours after ingestion. It should also be noted that the problem of undesirable flavours from feeding is often associated with sudden changes of diet -for example, the transition from dry winter feeds to rich green summer forage.

Direct absorption of substances by the milk following milking is less than formerly believed since, rather more volatile compounds pass through the animal. Recent studies have shown, however, that with whole milk highly volatile and fat-soluble organic solvents were directly and primarily retained by the fat phase, since, skimmed milk does not show this defect. The presence of these substances in premises for the collection, processing and storage of milk or milk products has often been the source of this occurrence and therefore their presence should be prohibited.

Another class of defects designated as "chemical flavours" can be caused by the contamination of milk by chemical agents included in the formulation of detergents and disinfectants used for cleaning recipients and equipment. For instance, alterations due to the presence of chlorinated and iodized compounds have been the most frequently reported.

Less frequently, phenol compounds from disinfectants and weed-killer have been considered responsible. This applies also to the development of foreign tastes and odours of the chlorophenol type transmitted by water disinfected with chlorine. It should also be mentioned that milk receptacles that have been used occasionally to contain or transfer other products (petroleum products, pesticides, non-food disinfectants and alcohol) can remain highly impregnated and, due to the residues present when reutilized normally, cause marked adverse changes in the flavour of the milk which can persist even after it has been greatly diluted.

Most technical treatments have little effect on the elimination of transmitted tastes and odours. Only vacuum treatment (degasing and evaporation) can decrease the extent of these defects, but their effectiveness is closely linked to the type of volatility of the offending substances in the physical conditions of the treatment applied.

TROUBLESHOOTING FOR BETTER-TASTING GOAT MILK

Goat milk is popular for many cheeses, and some people prefer it for other cultured milk products, such as yogurt, as well. The taste of goat milk is often a factor in making this decision. There are many factors that contribute to the taste of goat milk. Some people prefer the "goaty" flavor for which goat milk is well known. Others "put up" with the taste if that is their only option. Still others like their goat milk to taste just like cow milk. The issue is subjective and depends upon people's taste preferences.

Fresh goat and cow milk should taste basically the same. However, if goat milk has a "goaty" flavor, it can be attributed to a variety of causes: some harmless, some not so much, which may raise a red flag in terms of goat health, care, diet and environment.

STORE-BOUGHT GOAT MILK

Goat milk bought in a store often has a very goaty taste and has been processed differently than fresh, raw goat milk. While raw goat milk may be available in some stores, most often goat milk from the store has been pasteurized, a process in which live enzymes, bacteria, and nutrients are killed by heating the milk to certain temperatures. The freshness of the milk may also be compromised due to the time it takes to process, handle, package, and deliver the milk to its destination. There also may be levels of antibiotics, steroids, and medicines found in store-bought milk that could affect its flavor.

Goat Health

A possible health issue in the doe may cause the milk to taste "off" or goaty. Low-grade infections, illnesses, or viruses may affect the taste of the milk, and when treated, would restore a good flavor to the milk. A common condition affecting milk is called mastitis, an inflammation of the mammary gland (the udder) causing a chemical and physical reaction in milk. This condition is found more in large, mass-producing environments than in a smaller or family dairy environment. Mastitis has been proven to be associated with poor sanitation as well as trauma and bruising to the udder or teats. A clean living environment as well as healthy diet and immune system create a strong barrier against conditions such as mastitis. Changes in stress, extreme temperatures, poor living conditions and diet can contribute to taxing the health of the goat, possibly affecting the taste of her milk.

Care of the Animals

Poor sanitation is another piece in the goat milk taste puzzle. Ideally, goats should have enough space to eat, exercise, and explore. Proper feed, hay, and a good source of water should always be provided. Shelters should be kept with clean hay and regularly swept of manure and other organic refuse to minimize the contraction and spread of disease. These precautions help to keep goats healthy and provide a low-stress environment, all factors contributing to a flow of healthy and good tasting goat milk. Goats that have a specific, fixed diet of hay and feed will have fewer variables influencing the taste of their milk than those goats that are allowed to free range and graze upon seasonal growths (i.e., perennial grasses, wild garlic, onion, etc.) Certain grasses and wild growths can certainly affect the taste of the milk. The goat owner will need to monitor the diet of the animal and alter/experiment by reducing or increasing different feeds and grains in order to identify the possible effect on the taste of the milk.

Bucks (male goats) emit a strong musky odor, more prevalent during rutting, or mating, season. Female and castrated goats do not emit this odor. Many goat owners maintain that having a buck in habitation with the milking doe will cause the milk to have a goaty, musky taste. Other owners keep a buck in year-round with the does and declare no effect whatsoever on the taste of the milk. So this is a factor that may be considered in trying to make milk taste less goaty.

Milk Processing

The processing of the milk is one of the main factors affecting the taste of the milk. Goat milk contains the enzyme caproic acid, which causes it to turn "goaty" with age. So fresh milk properly processed is recommended for drinking and making dairy products without a goaty taste. Sanitation: It's important to take proper care of the milking utensils, and ensure the cleanliness of the goat udder and teats in order to reduce the possibility of bad bacteria influencing the taste of the milk.

Utensils must be cleaned and prepped before each milking session. The udder and teats should also be washed and dried with a clean cloth in order to clear away any hairs, dirt, straw, and possible feces that may have made contact since the last milking session.

Before collecting the milk during milking, a few sample squirts off to the side should be analyzed for any foul smell, thickness, or blood which could indicate some sort of illness or infection that may require attention in order to restore the milk to be fit for consumption and optimal taste.

Filtering: There will inevitably be some stray hairs or organic materials that fall into the pail during the milking session. Some people, in order to eliminate these materials from affecting the taste of the milk, may place a filter directly over the milking pail as another step of prevention during the milking process. Otherwise, filtering the milk after the session is complete is generally the accepted method.

Chilling: Once milking is complete, the milk must be filtered and chilled or frozen as quickly as possible to inhibit multiplication of bacteria, which can result in a more goaty taste. A milk filter can be anything from a coffee filter to cheesecloth or a clean, tightly-woven cloth to specific milk filters which may be purchased through a livestock supply outlet or catalog.

The longer the milk is left warm, the more goaty the taste will be. Some people desiring that taste in their cheeses may opt to make cheese with fresh goat milk that has not been cooled in order to take advantage of the natural process.

In warmer climates it is difficult to keep the milk from starting to turn goaty even before the milking session is over. Some have addressed this issue by chilling the bucket prior to milking or placing a pre-chilled object in the bucket during milking, enabling the chilling process to begin as soon as the milk comes out of the teat.

Genealogy

When all of the above has been tried, it may simply come down to the genetic makeup of the particular goat or goat line. Different does may have the exact same care and processing of their milk, yet the milk from one doe may taste goaty while the milk from another in the same herd does not.

TROUBLESHOOTING MILK FLAVOUR PROBLEMS

Sometimes certain nutrition programmes or management practices on the farm can cause off-flavour problems in milk. This can have long-term ramifications with consumers because of a poor tasting product. This can undermine consumer confidence in dairy products. Therefore, it is in everyone's interest to prevent these occurrences from happening regardless of the source. This fact sheet will address the common off-flavour problems and how they can be prevented.

CLASSIFICATION OF OFF-FLAVORS

Off-flavors commonly found in milk can be classified in three basic categories – the ABC's of off-flavour development.

1. Absorbed – feedy, barny, cowy, unclean, weedy, and musty.
2. Bacterial – acid, malty, unclean, fruity, and putrid.
3. Chemical – cowy (ketosis), rancid, oxidized, sunlight, and medicinal.

Absorbed flavour defects can develop before, during and after milking. It can occur when milk is left uncovered in the consumer's refrigerator or kept in cold rooms and dairy cases with other odour-producing foods.

Bacterial degradation results from bacteria that get into the milk upon contact with improperly washed or sanitized equipment, from external contamination, and is made worse by improper cooling. Infection of cows should not be considered as a source of high bacteria counts until all other causes have been eliminated.

Chemical defects can occur both before and after milking. The cowy or ketone flavour is the result of the animal suffering from ketosis. A foreign flavour can be caused by medications, a reaction to pesticides, disinfectants, or any number of contaminants. Rancidity and oxidation result from the degradation of milkfat.

RANCID FLAVOUR

A soapy-bitter taste is identifiable with rancidity. There appears to be a seasonal affect with the months between July and September having the highest occurrences. Rancidity is caused by a chemical development, which continues until the milk is pasteurized. It involves lipase and other enzymes, which react with the milkfat to form free fatty acids. The key to prevention is to have intact membranes around milkfat globules. Causes of rancid flavour are those things that result in weakened or broken milkfat globule membranes.

Some farm related causes include:

- Lack of adequate protein in the diet.
- Not feeding enough total energy for the level of milk produced.
- Milking cows longer than 305 days.
- Added stress when milking cows more than two times per day.
- Air leaks in pipeline milkers.
- Flooding of pipelines and receiver jars.
- Partial or less than every other day collection of milk from farms.
- Freezing in the bulk tank.
- Over agitation in the bulk tank.

Some non-farm related causes include:

1. Holding raw milk in processing plants more than 48 hours after collection.
2. Failure to empty and wash raw milk storage tanks every processing day.
3. Air leaks in pipes.
4. Running pumps in a starved condition.
5. Homogenization prior to pasteurization.

To correct a rancidity problem, start with the milk as processed and proceed back to the farm. Load samples of all milk received at the processing plant should be tasted regularly after laboratory pasteurization of 145°F for 30 minutes. Milk samples should be held for 48 hours after collection from farms and then tasted. Sometimes there is a slight butyric acid odour associated with rancid milk, but most of the time samples must be tasted to determine rancidity.

At the farm level, there are equipment observations that can be made. Check the pipeline and receiver jar during milking for foaming or flooding, or if the milk pump runs continuously. Check all fittings on pipeline milkers for tightness and close fit. Minimize air injection at claws and avoid over milking cows. No more than two units should be used on a 1.5-inch pipeline milker for each slope, i.e. two units on each side for a double slope line. Do not milk straight through weigh jars. Check the bulk tank for evidence of freezing or churning.

In herds with many late lactation animals, it may be necessary to sample milk from all four quarters on animals milking in excess of 305 days. The milk should be stored for 48 hours and then tasted. If milk samples taste rancid, some of their milk may need to be diverted from the bulk tank. Maintain milk quality by keeping the herd somatic cell counts below 250,000 per ml. Check the ration programme for proper fibre levels and particle size, fat content, and protein and energy levels appropriate for animal production.

Acid Degree Value (ADV)

Rancidity is characterised by the release of free fatty acids because of the action of the lipase enzyme. When consumers taste rancid milk, they are detecting the short chain fatty acids. An Acid Degree Value test can be used to measure the presence of long chain fatty acids. There is about a 70 percent correlation between flavour and ADV. ADVs are more meaningful for raw milk samples from individual farms. In mixed milk on trucks or in silo tanks the effect of milk from one farm is minimized. Rancid milk from one farm may cause a rancid flavour in the entire tank. Therefore, ADV results from mixed milk are not very meaningful. Milk samples from individual farms should have ADVs of less than 0.80. When results exceed 1.00, the milk will have a soapy-bitter taste. Some farms may have ADVs above 1.00 at the time of processing. ADVs should be conducted at the time the milk is processed rather than when the milk is received at the plant. When ADV results exceed 1.20, one can expect consumer complaints and a negative effect on consumption and sales. Action should be taken when samples results are above 1.00.

LACTOSE SENSITIVITY AND WHY MILK CAUSES GAS

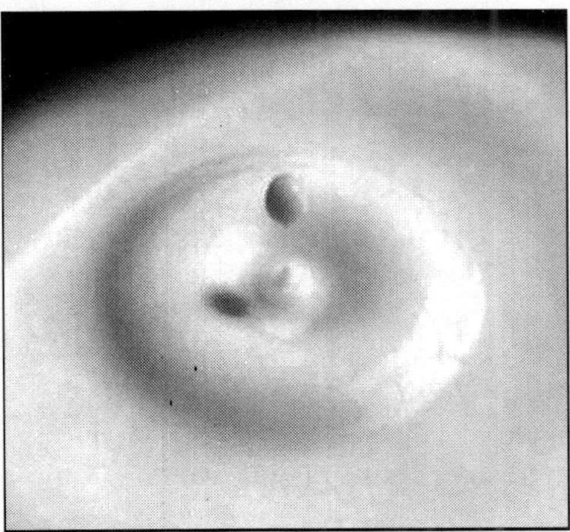

Lactose sensitivity is a very common cause of digestive problems. A person who is sensitive to the lactose in milk will often experience symptoms such as bloating, intestinal cramps and excessive gas after consuming dairy products. Ahead is what causes lactose sensitivity and intolerance, how to know if it's affecting you, what foods and drinks contain lactose and how to greatly reduces symptoms if you do have milk or one of the many other sources of lactose.

WHAT IS LACTOSE SENSITIVITY

Lactose is a type of sugar found in cow's milk and other milk products. It is made of a glucose and a galactose molecule. Lactose should be broken down in your upper intestine by the enzyme lactase. As infants, most of us produce a lot of lactase in our digestive systems and usually have no problems digesting milk.

As we get older though, usually around adolescence, the amount of the lactase enzyme that our bodies produce to break down lactose diminishes. This is quite normal, as milk is for infants and humans are the only animal that continues to try and drink it beyond infancy. Just how sensitive you are to lactose depends on just how much lactase enzyme your digestive system is still producing. As an example, some adults may be able to drink a glass of milk without any obvious side effects. Others may be able to have half a glass without discomfort, but any more than that will trigger symptoms.

For even more lactose sensitive people, even a splash of milk in their coffee can cause gastrointestinal distress. At this level, a person is considered lactose intolerant. Virtually any milk products or ingestion of lactose found in various other foods will quickly cause them problems. Someone who is very lactose intolerant usually knows it, as they experience bloating, stomach cramps and excessive flatulence soon after having milk or most other dairy products. More commonly, people have varying degrees of lactose sensitivity, making it more difficult to pinpoint milk, and its difficult to digest sugar, as the cause of their intestinal issues.

How Many People are Lactose Sensitive or Intolerant?

It's estimated that around one third of Americans would be recognized as lactose intolerant. This figure can be misleading though as it depends greatly on your ethnic background. Virtually all Native Americans and those of Asian descent have problems digesting the lactose in milk, around 75% of people of African and Caribbean descent have milk intolerance, and around half of Americans with Mediterranean heritage also show signs of lactose intolerance. Only Americans of northern European descent have commonly developed the genetic mutation to deal with lactose and can drink milk regularly without experiencing problems like bloating, abdominal pain and gas. It's estimated that only between 5% and 15% of these people would be considered truly lactose intolerant.

So while your background definitely plays a part in just how much dairy you can tolerate, the majority of adults do have a reduced ability to digest lactose. Unfortunately, despite all the potential digestive problems with lactose, food manufacturers seem to find milk sugar a cheap and useful ingredient and it is added to a wide variety of processed products as we'll see ahead. But first, why exactly does lactose cause gas?

Why Does Milk Make You Fart?

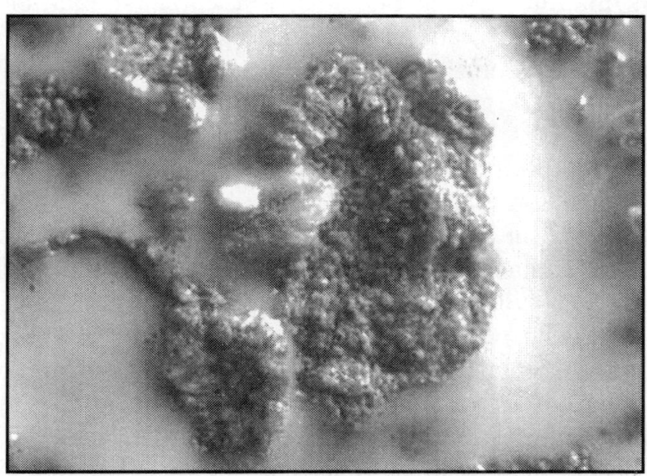

When your body isn't producing enough of the lactase enzyme in your small intestine, lactose in the food you eat or the products you drink like milk pass through to your large intestine undigested.

Certain bacteria in your colon love milk sugar and will quickly go to work fermenting it. This process creates a lot of gas and, due to the high levels of protein in milk, it is usually much smellier flatulence than that produced by the fermentation of the raffinose in beans. These extra gases in the gastrointestinal system can cause all sorts of problems and discomfort, from bloating and intestinal cramps and spasms, to diarrhea when pressure from flatus gas interferes with water absorption in the colon, and of course excessive flatulence.

In some respects, passing wind after drinking milk, and getting the gas out before it causes any more problems is the best thing that can happen. Not very comforting I know if it comes on during an office meeting or on a date. This process usually happens quite quickly and some individuals can experience bloating and flatulence after drinking milk in as little as 15 to 20 minutes. If they are going to have symptoms, most people will usually experience them within an hour of having milk if they are lactose sensitive.

Which makes you wonder why, if lactose is such an obvious cause of bloating and flatulence and it happens so quickly, why are milk products and other sources of lactose still so popular? Why haven't we worked out on a larger scale that milk makes you fart?

It's likely many people can get away with a small amount of milk, perhaps the amount they splash on their cereal in the morning (though surely nutritious hemp milk is a much better choice here). But then, with their lactase enzyme stores depleted, any of the many other sources of lactose we'll cover next, have the potential to cause gastrointestinal issues. Since food containing lactose travels through the digestive tract much slower than liquid, it may be many hours before symptoms appear. Someone who can have milk in the morning without obvious problems, but then ends up gassy in the evening from the lactose rich lunch he ate, may never make the connection between the two.

What Foods and Drinks Contain Lactose?

Here's a list of foods and drinks that contain lactose. You definitely want to avoid the products on this lactose list if you're intolerant. While milk is the most obvious source of lactose, there are many other potential culprits of flatulence and other intestinal problems caused by undigested milk sugar. The majority of us though, with varying levels of lactose and milk sensitivity, would also do well to reduce our consumption of the following where possible to prevent bloating, abdominal cramps, flatulence and other symptoms caused by not breaking down the lactose sugar in milk.

- All dairy milk, especially skim milk which has even higher levels of milk sugar, but all milk derived from cows or goats will contain lactose in some of the highest levels.
- Protein powders made with whey can be a concentrated source of lactose and cause flatulence problems for many people.
- Butter and all forms of cream contain less lactose than milk, but still enough to cause problems for the lactose intolerant or highly sensitive.
- Cheese will generally contain less lactose the harder it is. So hard aged cheeses like Romano, parmesan or cheddar will have less milk sugar than softer cheeses like Camembert or mozzarella, which are generally higher in lactose.
- Ice cream can unfortunately be particularly bad. Not only is it made from milk, but manufacturers have incentive to sweeten it even further with added milk sugar.
- Yogurt, in theory should be lower in lactose as the beneficial bacteria predigests the milk sugar. Unfortunately, what passes for yogurt in the supermarket fridges these days is usually nothing more than sweetened, thickened milk with very little probiotic content. A good quality and unflavored traditional Greek or cultured yogurt will likely be much lower in lactose, but these can be difficult to find.
- Canned soup is a less obvious source of lactose, but often quite a high one. Check the labels for any milk-based ingredients, but generally any cream-based soup will be full of lactose.
- Savory snacks often contain milk powder which again contains lactose and anything cheese flavored will usually be especially high in flatulence causing milk sugar.
- Sweets like milk chocolate (dark chocolate is much lower and

healthier), cakes, puddings, cookies and particularly doughnuts are all potentially significant sources of lactose.

- Sauces, gravies and salad dressings like mayonnaise can all contain lactose. You can check the labels for milk solids, milk powder, anything starting with whey and of course lactose itself.
- Breakfast cereals are also sometimes made with milk powder or milk solids, though of course this pales into insignificance if you're already eating them with a big bowl of cow's milk.
- Bread is another commonly eaten food that can contain lactose. The amount on its own may not cause problems, but as you can see with all the different products containing milk sugar, it's all the more to ferment in the lower intestine.
- Even processed meats like sausages and luncheon meat can have lactose in them, as can breaded or battered meats, and are worth avoiding for the lactose intolerant. Of course pizza with its combination of soft cheese, bread and often processed meats will be a triple whammy and usually high on the lactose scale.
- Even birth control pills, headache tablets and some drugs and supplements contain lactose as an ingredient for some strange reason. Alone they are unlikely to have an effect, but if your body cannot produce any more lactase enzyme then it all adds up.

As you can see it's a fairly extensive list of products with lactose in them. If you're looking on food labels trying to avoid milk sugar you want to watch out for anything that says: milk powder, milk solids, milk protein, whey solids or protein, nonfat dried milk, casein, sodium caseinate and obviously lactose itself.

LACTOSE INTOLERANT VERSUS LACTOSE SENSITIVE

For someone who is truly lactose intolerant it's important to avoid as many foods as possible on the above list to prevent gastrointestinal problems. It's much more serious than just cramps, bloating and flatulence. Constantly dealing with the fermentation of lactose can leave your digestive system in a weakened state and lead to harmful bacterial overgrowth and lowered immunity. For a person who is only slightly lactose sensitive, just avoiding milk itself may be enough to prevent all but occasional problems. This hemp milk is a great alternative, high in important omega-3 fats and completely lactose-free. Those with moderate lactose and milk sensitivity (and I'd include myself in this group) would want to avoid milk and generally some of the higher lactose containing foods like soft cheese, ice cream, supermarket yogurt, snacks with milk solids, creamy soups, pizza, doughnuts and milk-based sauces and gravies. If you'd like to check for sure if you are lactose intolerant or on the higher scale of lactose sensitive then there is a simple test available from your GP called the 'Lactose Intolerance Breath Test'.

HOW TO DRINK MILK WITHOUT FARTING

If you really can't bear the thought of a life without flavored milk, ice cream and Camembert on crackers, there are lactase supplements that you can take to replace the milk sugar digestive enzymes your body is no longer making in large enough quantities. Lactaid is the most well known of these, but these ones contain the same 9000 FCC (Food Chemical Codex) units of lactase per capsule as the highest strength Lactaid. At 180 capsules versus the usual 60 in Lactaid they are also about half the price and work just as well for me. Many in the reviews seem to agree.

I take them at the same time as an occasional ice cream or pizza and there's none of the usual bloating, tummy rumbling or gas. It's a good idea to take one of the silver blister packs with you if you think you could be having foods or drinks containing a lot of lactose and have them either just before or with your meal or milk based drink.

Beware of other cheaper brands of lactase enzyme caplets or chewables that only contain 1000 or 2000 FCC units. This won't be enough for all but the lowest of lactose sensitivity and definitely not enough if you are lactose intolerant. Millions of people in America have problems with the lactose in milk but don't know it and experience abdominal bloating and cramps, excessive flatulence and other gastrointestinal problems regularly. Please share this page with your friends and let people know there is a solution.

UNDERSTANDING THE CAUSES AND IMPORTANCE OF MILK FAT AND PROTEIN VARIATION

Most milk marketing orders in the U.S. now employ a multiple component pricing system that pays producers on the basis of milk fat, true protein, and other dairy solids. This new pricing method derives component values from prices for manufactured dairy products (cheese, butter, nonfat dry milk, and dry whey), which rise and fall with changing market conditions. As a result, milk component levels have taken on new importance in herd management. In addition to being indicators of cow health and nutrition, component levels now directly impact farm income. This factsheet will describe the variation found in production of milk components, factors that contribute to this variation, and strategies to improve component production. Generally, fat and protein content of milk are positively correlated within a population of dairy cattle; however, different breeds of cattle vary in average component levels. Holsteins have the lowest fat and protein content, while Jersey and Guernsey breeds have the highest. Because Holsteins produce more milk, they generally have a higher total yield of fat and protein than other breeds.

Table. Average Fat and Protein Content of Milk Produced by Different Breeds. [1]Ratio of fat to Protein.

Breed	% Fat	% Protein	F:P[1]
Ayrshire	3.86	3.18	1.21
Brown Swiss	4.04	3.38	1.20
Guernsey	4.51	3.37	1.34
Holstein	3.65	3.06	1.19
Jersey	4.60	3.59	1.28

Production of milk fat and protein can vary tremendously from one herd to another. A recent summary of milk shipped in the Mideast federal order from 2000 through 2002 showed that herd average milk protein ranged from 1.57% to 4.66%, with an average of 3.05%. Milk fat ranged from 1.77% to 5.98%, with an average of 3.76%.

These data represent real herds and actual production. While these data do not provide any information about breed, the range in herd performance is huge. The same study, but presents the range that includes 68% of the population. Even when the extremes are eliminated, this study indicates that many herds are producing components below average for their market and their breed. This presents an opportunity to improve component production and income from milk sales.

Some comparisons of the value of milk sold as the component levels change. If a Holstein herd currently produces milk with 3.5% fat and 2.9% protein, that herd is below the breed average for components. In addition, if that herd ships milk in the Mideast federal order, it is also below the market average for component levels. Increasing fat and protein to breed average would increase the value of milk sold by 58 cents per hundredweight. Increasing to market average would gain the herd 72 cents per hundredweight. If the herd could increase component percentages by 0.32 for fat and 0.19 for protein, an extra 90 cents per hundredweight would be generated.

Table. Comparison of Gross Milk Price at Various Levels of Fat and Protein Production.

	Fat, %	Protein, %	Gross Price, $/cwt	Difference[1], $/cwt
Current test	3.50	2.90	13.41	0.00
Holstein average	3.65	3.06	13.99	-0.58
Mideast average[2]	3.76	3.05	14.13	-0.72
Increase by 0.5 SD3	3.66	2.98	13.83	-0.42
Increase by 1 SD3	3.82	3.09	14.31	-0.90

Assumptions: herd SCC of 150,000 and 5.65% other solids for each case. Component prices assumed are averages from Jan. 2000 to June 2005 for the Mideast federal order: fat $1.5256/lb, protein $2.1700/lb, other solids $0.0691/lb, SCC adjustment rate $0.00068. Producer price differential for each case is $1.25/cwt, an average received since, 2000 for milk shipped to processing plants located in western Pennsylvania. [1]Current gross price minus the gross price for each scenario. [2]Data from summary of milk shipped in the Mideast federal order from 2000 through 2002. One standard deviation (SD) is 0.32% for fat and 0.19% for protein.

Increases of this magnitude are possible with changes in nutrition and management. What's more, if nutritional changes are implemented it is very likely that milk production will increase along with component levels, which would add even more income. We have developed a simple spreadsheet to help you calculate the gross milk price for various component levels.

FACTORS AFFECTING MILK COMPOSITION

There are many factors that can affect milk fat and protein, and many of them can be manipulated to enable you to achieve higher than average levels of milk components. Keep in mind that herds that are below breed average will have more opportunity to improve component levels. Herds that are already above average may have better success by focusing on increasing milk yield, which will increase the total amount of fat and protein produced.

Factors Other Than Nutrition

Stage of lactation affects milk protein and fat percentages very similarly. The highest amount of protein and fat in milk is found just after freshening, in colostrum. Levels drop to their lowest point between 25 and 50 days after calving and peak at 250 days as milk production begins to decrease. Age tends to cause both milk fat and protein to decline as the animal becomes older. Milk fat falls about 0.2% each year from the first to fifth lactation likely as a result of higher production and more udder infections. Protein decreases 0.02 to 0.05% each lactation as animals age. Season dramatically affects milk fat and protein. The hot, humid months (July and August in the NE) depress fat and protein content.

There is a gradual increase of protein and fat in milk through the fall and peak levels occur in the colder months of winter. As temperatures increase through the spring, component levels are gradually decreased. These changes may be indicative of feed intake patterns, which are lower in summer due to changes in weather and temperature.

Mastitis infections reduce fat and casein but increase blood protein content of milk. Somatic cell count (SCC) also is elevated during mastitis. Herds that have continuous mastitis and SCC problems take a double or triple hit on milk price. The component value is reduced, plus in some federal orders there is a deduction for SCC over 350,000. Quality premiums from the milk handler may be lost as well.

Milk fat and protein depression also can occur from mechanical errors, such as cooling problems in the bulk tank, sampling problems, and over agitation in the pipeline. Genetics and inheritance account for 55% of the difference between cows in protein and fat content of milk. The heritability estimates for milk and its components. Heritability indicates the proportion of observed differences that are due to genetics, while the reciprocal is assumed to be due to environmental factors. Protein and fat percentages are more highly heritable than yield of milk and components. Milk yield is positively correlated to yield of fat and protein; however, milk yield is negatively correlated to fat and protein percent. For many years, sires have been selected for high yields of milk, which has resulted in very slow increases in fat and protein percentages over time. Herds that are more than one standard deviation below the breed average for fat or protein may benefit from including component yields in sire selection criteria. However, because fat and protein percentages are negatively related to milk yield, changes in herd component percentages are not likely to be achieved through genetic selection alone.

Table. Heritability (h^2) Estimates for Milk and its Components.

	Holstein		Jersey	
Trait	h^2	SD[1]	h^2	SD[1]
Fat, %	0.58	0.23	0.55	0.28
Protein, %	0.51	0.14	0.55	0.20
Fat, lb	0.30	52	0.35	50
Protein, lb	0.30	37	0.35	36
Milk, lb	0.30	1444	0.35	1204

Nutritional Factors and Feeding Practices

Of all the factors affecting milk composition, nutrition and feeding practices are most likely to cause problems; however, management changes made here are able to quickly and dramatically alter production of fat and protein. Milk fat depression can be alleviated within 7 to 21 days by changing the diet. Milk protein changes may take 3 to 6 weeks or longer if the problem has been going on for a prolonged period. Nutrition or ration formulation changes are more strongly correlated to milk fat content than milk protein. Milk fat can be changed by 0.1 to 1.0 percentage points, while protein is seldom altered more than 0.1 to 0.4 points by nutritional changes. For these reasons, nutrition and feeding management are considered the best solutions to a milk fat or protein problem other than genetics.

Source of Milk Components. Digestion of fibre in the rumen produces the volatile fatty acids (VFA) acetate and butyrate. Butyrate provides energy for the rumen wall, and much of it is converted to beta-hydroxybutyrate in the rumen wall tissue. About half of the fat in milk is synthesized in the udder from acetate and beta-hydroxybutyrate. The other half of milk fat is transported from the pool of fatty acids circulating in the blood. These can originate from body fat mobilisation, absorption from the diet, or from fats metabolized in the liver.

Rumen microbes convert dietary protein into microbial protein, which is a primary source of essential amino acids for the cow. These amino acids are used by the mammary gland to synthesize milk proteins. Glucose is required to provide energy to support this protein synthesis. Glucose is either formed from the VFA propionate in the liver, or absorbed directly from the small intestine. If too little propionate is absorbed from the rumen, the cow will have to breakdown amino acids and convert them to glucose (a process called gluconeogenesis); this can reduce the supply of amino acids available to make milk protein. In addition, some albumin and immunoglobulin protein is transferred directly to milk from the blood.

Rumen Function. The relative amounts of protein and energy that are available in the rumen at a given time is the major factor affecting rumen fermentation and therefore milk components. Any diet or management factors that affect rumen fermentation can change milk fat and protein levels. Consistently providing adequate energy and protein and balanced amounts of rapidly fermentable carbohydrate and effective fibre are keys to maintaining optimum levels of milk components. The challenge in feeding for milk components is that high energy, low fibre diets that increase milk protein are likely to reduce fat levels. This may also be the case in some diets with rumen modifiers, such as Rumensin®; however, this product has other ways to affect the rumen that do not necessarily alter milk components.

Feeding Management. Any situation that causes cows to eat abnormally or limits feed intake may affect milk components. Examples include: overcrowding at feed bunks, housing heifers with older cows in facilities at or near full capacity, feeding rations that encourage sorting, feeding infrequently in a conventional system (non-TMR), failing to push feed up or feed TMR often enough, feeding protein feeds before energy feeds and feeding grain before forage in non-TMR systems. These conditions can create slug feeding (one or two meals per day versus 10 to 15) or allow cows to eat high grain meals part of the time and high forage meals the remainder of the day. Ensure that fresh feed is available 20 hours each day, spoiled feed is removed from bunks, and shade or cooling is provided during hot weather to help maintain normal intake and normal meal patterns. Poor ventilation or cow comfort also can depress milk fat and protein production by reducing intake. Finally, make ration changes gradually to allow rumen microorganisms time to adapt. Any reduction in rumen microbial protein production from nutrition or feeding management imbalances will reduce milk protein by way of less microbial protein for the cow to digest and depress fat by limiting VFA production in the rumen.

Body Condition. Proper body condition is essential so that high producing cows can draw on body stores of nutrients to support milk production. If body stores are minimal, yields of milk and milk components will suffer. On the other hand, excessive body condition increases the risk of metabolic problems and calving difficulty. Weight loss in early lactation can increase milk fat content for a short period of time. Both thin and fat cows tend to have low milk fat in later lactation. Protein can be depressed at calving if animals are overly obese or underweight. In addition, some research shows that underfeeding protein during the last three weeks before calving can depress milk protein.

Energy Effects. In general, as energy intake or ration energy density increase and/or fibre decreases, milk fat content will be reduced, while protein is increased. In contrast, as ration fibre levels increase and/or energy is reduced, milk protein is depressed and milk fat is increased. Lack of energy intake or lower ration digestibility may reduce milk protein by 0.1 to 0.4%. This reduction may result from underfeeding concentrates, low forage intake, poor quality forage, failure to balance the ration for protein and minerals, or inadequately ground or prepared grains. Shifting rumen fermentation so that more propionic acid is produced is apt to increase milk protein and decrease fat content. However, excessive energy intake, such as overfeeding concentrate, may reduce milk fat content and increase milk protein. Normal protein levels can be expected when energy needs are being met for most of the cows. Often this is impossible to achieve with high producing animals.

Protein Effects. A deficiency of crude protein in the ration may depress protein in milk; marginal deficiency could result in a reduction of 0.0 to 0.2%, while more severe restriction of diet crude protein would have greater impact. However, feeding excessive dietary protein does not increase milk protein, as most of the excess is excreted. Dietary protein has little effect on milk fat levels within normal ranges.

Diet protein type also could affect milk protein levels. Use of non-protein nitrogen (NPN) compounds, like urea, as protein substitutes will reduce protein in milk by 0.1 to 0.3% if the NPN is a main provider of crude protein equivalent. Rations higher than recommended in soluble protein may lower milk protein by 0.1 to 0.2 points. NPN levels in milk will be increased by excessive protein or NPN intake, heavy feeding of ensiled forages, ensiled grains, immature pasture and lack of rumen undegradable protein in the diet. Balance rations for crude protein, rumen undegradable protein, rumen degradable protein, and soluble protein. For high producing cows, balancing for amino acids also may be required.

Concentrate Intake. An increase in the intake of concentrates causes a decrease in fibre digestion and acetic acid production. This creates an increase of propionic acid production. Propionic acid production encourages a fattening metabolism that is in opposition to milk fat. Addition of buffers to some rations may help to prevent acidosis; this will not change milk protein, but will increase milk fat content. Animals that eat a substantial amount of concentrates or a low ratio of dietary forage to concentrate may develop acidosis even when buffers are added to the ration.

The nonfiber carbohydrate (NFC) portion of the diet is highly digestible and can influence both fat and protein in milk. Excessive amounts of NFC can depress fibre digestibility, which reduces the production of acetate and leads to low milk fat (1% or more reduction). At the same time, greater propionate production allows higher milk protein levels of 0.2 to 0.3%. Generally an NFC of 32 to 38% of ration dry matter is recommended to optimize production of milk fat and protein.

Forage Level and Physical Form. Balance rations for lactating cows to contain at least 40 to 45% of ration dry matter from forage. This may be altered by the level of corn silage in the ration and the level of high-fibre by-product feeds in the ration. Low forage intake can cause a major reduction in the fat content of milk due to low fibre levels. Several potential reasons for low forage intake are inadequate forage feeding, poor quality forage, and low neutral detergent fibre (NDF) content in forage that was cut too young or late in the fall. Target a forage NDF intake of 0.9% of bodyweight daily. Although low forage (high energy) diets increase milk protein production, this strategy is not recommended. The low forage levels contribute to acidosis and laminitis; they do not promote good health for the rumen or the cow in the long run.

Protein and fat content also can be changed due to the physical form of forage being fed. Much of this is related to ration sorting and failure to provide a consistent diet throughout the day. Coarsely chopped silage and dry hay are the most common causes of sorting. At the other extreme, very finely ground diets negatively affect rumen metabolism and depress fat and protein production. Monitor ration particle size to ensure that adequate effective fibre is provided, TMRs are mixed properly, rations are distributed evenly to all cows, and sorting is minimal.

Added Fat or Oil. Adding fat to the ration can affect milk component levels depending on the amount and source of fat. Fat is generally toxic to rumen microbes and may reduce fibre digestibility when fat from natural sources exceeds 5% of ration dry matter. If rumen inert or bypass fat is used, total fat content may safely reach 6 to 7%. At low levels of dietary fat, milk fat content could increase slightly or show no change at all. Milk fat is reduced at higher levels, especially with polyunsaturated oils. If fat or oil is rancid, milk fat content decreases even at low levels of consumption. Milk protein content may be decreased by 0.1 to 0.3% in high-fat diets. This may occur due to reduced blood glucose levels.

EXTREMELY HIGH MILK FAT

High milk fat content often occurs in herds that are off in feed and may have ketosis problems. Percent fat may be reduced for sick animals, but total fat may be higher for the herd. This may occur in herds fed large amounts of good quality forage combined with moderate concentrate levels. Producing an abnormally high level of fat is not economically feasible, because it usually indicates that total milk production is low. Herds that depend primarily on milk income would be better served to increase total milk yield and keep fat percentage somewhat below the attainable maximum. Herds with unusually high milk fat are encouraged to reduce forage intake if it is on the high side, increase concentrate feeding, and manage the nutrition of dry and transition cows more closely to control problems with low intakes and ketosis.

MAINTAINING COMPONENT LEVELS

Recommendations for normal fat and protein content can be achieved by feeding a balanced ration that meets the chemical and physical needs of the cow. Key management practices to accomplish this goal include: regular forage tests for energy, minerals and protein; regular tests of TMR and concentrates to see if they meet herd requirements; evaluation of forage and TMR particle size; use of production records to track component yields; and use of body condition scoring to evaluate the success of nutritional programmes.

Monitor milk component percentages from your milk handler or DHI records by month. Over time this will allow you to develop a normal range. If components suddenly drop out of their normal range, investigate and find the cause. Paying close attention to component levels may also allow you to make nutritional changes in response to market conditions. Any decisions of this kind should be analysed to determine the additional cost compared to the additional revenue.

3

The Milk Taste Test and Microorganism

Milk taken from a healthy goat in hygienic surroundings and stored correctly in sterile containers should be sweet smelling and free from any noticeable flavour. A good indication of taint-free milk is when your visitors need to enquire whether or not you've served them cows' or goats' milk on their breakfast cereal or in their tea.

HUMAN HEALTH CONCERNS

It's not for no reason that dairy products attract so much attention from the environmental health department and other food agencies. Milk is an ideal medium for growing bacteria and absorbing taints from extraneous sources. Poor dairy hygiene can lead to nasty-tasting milk but it also has the potential to cause illness when specific bacteria are present. Bacteria reproduce at an alarming rate.

Correct cooling not only slows down the growth of bacteria but also inhibits the activity of the enzyme lipase. Lipase will do you no harm but it's responsible for imparting a 'goaty' taint to the milk. There seems to be a small minority of consumers who actually relish this flavour in their goats' milk! I came across an article on the internet, written by a commercial cheese maker, extolling the presence of high levels of lipase which, said the writer, "gave goat cheese its distinctive flavour". This may be acceptable, and even welcome, to the cheese connoisseur but, to most families, the presence of any noticeable goatiness is a real turn-off.

Anyone choosing to sell goats' milk must be licensed and their milk must undergo regular microbiological tests to ensure it's not harmful to health. Supermarkets will also demand, as an extra measure, that it's heat treated, to ensure all bacteria are killed off and homogenised to ensure it has a shelf-life in excess of 10 days. This will affect the flavour. I've purchased supermarket goats' milk with almost no distinguishable flavour but, on other occasions, the same brand has been unmistakably goats' milk – not unpleasant, just different to cows' milk.

Lipolysis

Lipolysis, the release of excessive amounts of fatty acids into the milk, is responsible for the distinctive goaty flavour of some milk. Poor dairy hygiene and careless milking practices often shoulder the blame for goaty tasting milk but this is only partly true. The characteristic goaty flavour only becomes apparent with over activity of the lipase enzyme. Because the enzyme works with bacteria, a high bacteria count in the milk will exacerbate the situation but even the most hygienic milking regime can't always prevent it.

Bacteria may be present as a result of sub-clinical mastitis. Hormonal activity at oestrus and after kidding is often responsible for brief periods of tainted milk in some goats, and subjecting goats' milk to excessive agitation will damage the fragile cells and also create this fatty acid taint. High lipase activity is often particular to an individual goat and generally only in the circumstances.

However, there are goats that manifest high levels of lipase at all times of their lactation. These goats are fortunately rare, but if you have such a goat, please don't damage the reputation of goats' milk by offering her milk for general consumption. Lipase cannot survive at a temperature of 56°C, therefore heating the milk to this temperature immediately after milking will destroy the enzyme and prevent any taint developing. This is a good way of testing for this fatty acid taint. If you have a goat with high lipase activity, pasteurising her milk as soon as possible after milking could solve the problem.

External contamination

Contamination of milk from external sources is usually fairly easy to identify and is invariably the result of careless storage. Never recycle plastic containers. Those that have held fruit juice can impart strange fruity flavours to the milk. Any that have previously held milk are notoriously difficult to sterilise because plastic has a softer, less robust, surface than glass or stainless steel. Containers that have held branded spring water can be used once but are then best discarded.

Milk stored in the fridge should always be in a sealed container. Some excellent glass jugs are available, designed to stand in the door of the fridge, and they come complete with a plastic lid. As the milk isn't in direct contact with the plastic, though, it doesn't pose the same threat as a plastic jug or a recycled plastic bottle.

An unpleasant metallic taint in milk is the result of chemical traces left on equipment due to inefficient rinsing. Oxidation produces a cardboard flavour in the milk. It's not a common cause of milk taint but will occur in milk left uncovered for any length of time in sunlight or under fluorescent light.

Food

Taints that occur in very fresh milk fall into two categories. The most straightforward and possibly the easiest to diagnose are the transient taints connected with the consumption of strong flavoured foods. Large quantities of swede, cabbage or kale need the maximum time for digestion prior to milking. They are best presented immediately after a milking. Sugar beet pulp is a welcome and palatable addition to a goat's diet, especially during winter but, if it's fed to excess, it will impart a fishy taint to the milk.

It's difficult to control what a goat chooses to eat while she's free ranging but being aware that several plant species can be problematic will help to manage the situation. There are several items listed under the heading of Plants that can taint milk in HMSO Book of Poisonous Plants. Some, for example hemlock, yew, horsetail, bracken, buttercup, henbane and laburnum are poisonous to a greater or lesser degree, and if eaten in any quantity, the goat isn't likely to live long enough for milk taint to be a problem!

Other plants known to taint milk include wild garlic, elderflower, mint, fools' parsley, birdsfoot trefoil, shepherd's purse, ivy leaves (a useful tonic in small doses but beware of the poisonous berries), oak, cress, sweet clover (melilotus), wood sorrel and various plants from the compositae family. Apart from those known to be poisonous, there's no need to ban these plants from a goat's diet, but if you notice off flavours in the milk that can't be explained any other way, a walk around the pasture observing what the goats are eating may give you an answer. It's also worth noting that carrots and peas sweeten the milk.

Goat health

The second category of tainted fresh milk involves the health of the goat and is a much more complicated issue. Mastitis, metabolic disorders, worms and mineral deficiencies all have an effect on milk production and flavour. Mastitis is usually accompanied by physical changes in the milk such as clots and stringiness. In severe cases the goat will appear ill and be running a high temperature. Her milk looks distasteful and is obviously unfit for consumption.

Goats suffering from sub-clinical mastitis will show no symptoms other than a lowered milk yield. The milk will look normal but the taste test will reveal a slight saltiness. This must be investigated as the goat could be shedding harmful bacteria into her milk. Ketosis or acetonaemia is caused when the goat can't consume sufficient energy foods for her needs and is unable to digest protein efficiently. Acute cases are recognised by the smell of keytones present in the milk and often on her breath – the smell is distinctive, of pear drops or nail varnish. The milk will also taste strangely sweet. The condition occurs shortly after kidding and is accompanied by a reluctance to eat concentrate at the very time she should be having maximum rations. Ketosis is a life threatening metabolic disorder caused because the goat isn't eating and is living off her body reserves. It won't get better by itself but responds quite quickly when treated with drenches available from the vet.

Flavour

Sometimes it's difficult to describe an off-flavour. Is it bitter? Does it lack sweetness? Does it leave an aftertaste? When I kept goats commercially, my taste test awarded marks out of 10. I awarded full marks for milk that tasted wholesome, creamy and sweet, that I knew would be enjoyed by small children. It was rare for any goat to achieve less than eight.

If any goat did have a problem, it occurred soon after kidding and, in my experience, was more likely to be related to its metabolism rather than its hormones. Vitamin B12 is produced in the rumen. A deficiency affects the goat's metabolism, and according to the Goat Veterinary Society, is the most frequent cause of tainted goats' milk. Normal healthy goats receiving a well-balanced and mineral rich diet shouldn't suffer any deficiency but there are certain conditions, notably infection from worms and shortage of cobalt, which will upset this delicate balance and hinder the production of vitamin B12.

Bitter-tasting milk is often blamed directly on a shortage of the trace element cobalt. The full story is that cobalt is essential for the production of B12 and as it's not stored in the goat's system, it must be available in the goat's diet on a daily basis. A diet containing a wide variety of plant species, either fresh or as hay, and ideally from a herbal ley containing chicory, together with the provision of a mineral supplement formulated for goats and a cobaltised mineral lick should ensure there is no shortage of cobalt.

Worm control is an essential aspect of goat husbandry and even a relatively mild worm infection will inhibit the production of B12. Therefore, if your goat is giving tainted milk, check your worming routine and arrange a faecal egg count.

Management

Lack of published research means that goat keepers often have to try and work out their own management strategies. Some goat keepers routinely inject newly-kidded goats with vitamin B12. Others maintain that tempting their goat with a Marmite sandwich gives good results. I've heard of the problem of bitter tasting milk being resolved by giving a goat proprietary antacid tablets for a few days. The calcium carbonate appears to address the calcium to phosphorus ratio that can be upset following kidding. A diet high in roughage but with insufficient energy foods to meet the goats' requirements will result in strong-tasting milk. This often occurs after kidding and rights itself after a week or so but should be viewed as a warning that diet management during and after kidding wasn't perfectly suited to the goat's needs.

All off-flavours in milk have a cause and should be investigated and, if necessary, corrected. If the milk isn't destined for human consumption, it doesn't really matter if it tastes of wild garlic or has been left to oxidise under a fluorescent light. If lack of vitamin B12 or an inappropriate diet is the cause, the taint is a warning that all is not right with the goat and her long term health could be jeopardised.

MICROBIOLOGY

Micro-organism is the term applied to all microscopically small living organisms. We tend to associate micro-organisms with disease. Micro-organisms which cause disease are called pathogens. However, few micro-organisms are pathogens and micro-organisms play a crucial part in the life of our planet. For example, they provide food for fish, they occur in soil where they provide nutrients for plants and they play an important role in ruminant digestion.

In dairying some micro-organisms are harmful—e.g. spoilage organisms, pathogens—while others are beneficial—cheese and yoghurt starters, yeasts and moulds used in controlled fermentations in milk processing. The micro-organisms principally encountered in the dairy industry are bacteria, yeasts, moulds, and viruses.

BACTERIA

Bacteria are single-celled organisms. They are present in air, water and on most solid materials. Bacterial cells are very small and can only be seen with the aid of a microscope. When observed under a microscope the cells can be seen to differ in shape and in conformation of groups of cells. Cells are either spherical or rod-shaped. Spherical bacteria are called cocci; those that are rod shaped are called bacilli. This is the first basis for differentiating between bacterial cells.

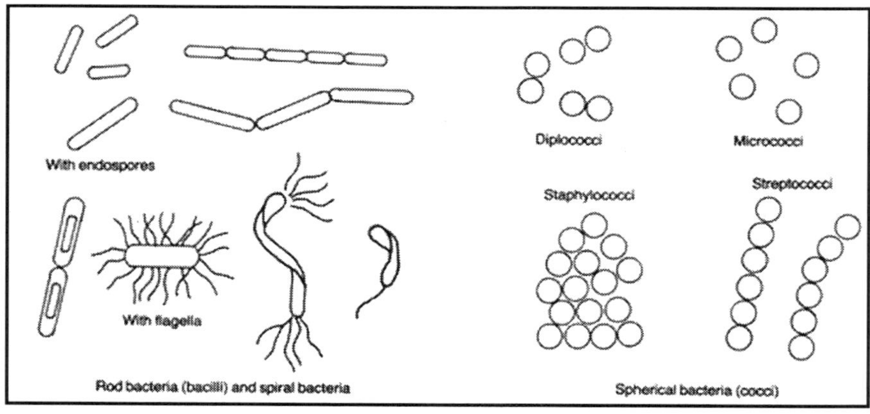

Fig. Rod-shaped (Bacilli) and Spherical (Cocci) Bacteria.

Bacteria are also classified according to cell cluster formation:

Diplococci – two cocci cells paired

Staphylococci – a number of cells clustered together

Streptococci – a number of cells arranged in a chain

Some bacteria are capable of locomotion by means of flagellae—long, hair-like appendages growing out of the cell. Some rod-shaped bacteria contain spores. These are formed when the cells are faced with adverse conditions, such as high temperature: once suitable conditions are reestablished the spores germinate to form new cells.

Close examination of the simple cell reveals that it is composed of the following components:

- Cell wall—this gives the cell its shape and retains the constituents;
- Cell membrane—used for filtering in food constituents and discharging waste products;
- Nucleus—where the genetic material of the cell is stored;
- Cytoplasm—a semiliquid proteinaceous substance which contains starch, fat and enzymes.

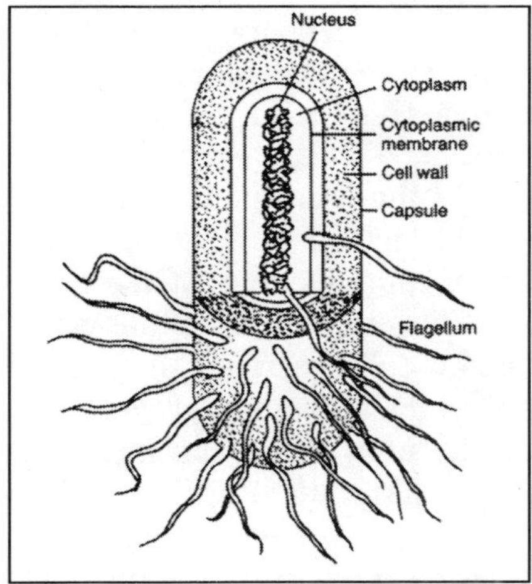

Fig. Schematic Illustration of Bacterial Structure.

The cell membrane is semipermeable and allows the cell to feed by osmosis, i.e. the exchange of water between the cytoplasm of a living cell and the surrounding watery material. Only small molecules can pass in and out of the cell, e.g. with a sugar solution on one side of a semipermeable membrane and water on the other, water will diffuse in, diluting the sugar solution.

The sugar molecules cannot pass out so a hydrostatic pressure, known as osmotic pressure, develops. Bacteria can feed by selective intake of nutrients dissolved in water. They can also take in nutrients against the normal osmotic flow—active transport.

BACTERIAL GROWTH

Bacterial growth refers to an increase in cell numbers rather than an increase in cell size. The process by which bacterial cells divide to reproduce themselves is known as binary transverse fission. The time taken from cell formation to cell division is called the generation time. The generation time can therefore be defined as the time taken for the cell count to double. The of bacterial growth following inoculation of bacteria into a new growth medium.

The following phases can be identified:

1. *Lag phase:* There is usually some delay in growth following inoculation of bacteria into a new medium, during which time the bacteria adapt to the medium and synthesise the enzymes needed to break down the substances in the growth medium.
2. *Log phase:* Once the bacteria have adapted to the new medium they start to reproduce quickly and their numbers multiply evenly for each increment of time. A plot of the log number of cells against time gives a linear relationship: this is therefore called the log phase. The cells are at their greatest activity in this phase. Transferring cultures to a fresh medium at regular intervals can maintain the cells in an active state. An active culture can rapidly dominate any new environment.
3. *Stationary phase:* As the bacteria dominate the growth medium, they deplete the available nutrients or toxic waste products accumulate, slowing the rate of reproduction. At the same time, cells are dying off: A state of equilibrium is reached between the death of old cells and formation of new cells, resulting in no net change in cell numbers. This phase is called the stationary phase.
4. *Death phase:* In the next phase the formation of new cells ceases and the existing cells gradually die off: This is called the death phase.
5. The log phase can be prolonged by removing toxic waste, by adding more nutrients or both.

The log phase can be prolonged by removing toxic waste, by adding more nutrients or both. The log phase can be prolonged by removing toxic waste, by adding more nutrients or both.

Factors affecting bacterial growth

Bacterial growth is affected by (1) temperature, (2) nutrient availability, (3) water supply, (4) oxygen supply, and (5) acidity of the medium.

Temperature: Theoretically, bacteria can grow at all temperatures between the freezing point of water and the temperature at which protein or protoplasm coagulates. Somewhere between these maximum and minimum points lies the optimum temperature at which the bacteria grow best. Temperatures below the minimum stop bacterial growth but do not kill the organism. However, if the temperature is raised above the maximum, bacteria are soon killed. Most cells die after exposure to heat treatments in the order of 70°C for 15 seconds, although spore-forming organisms require more severe heat treatment, e.g. live steam at 120°C for 30 minutes.

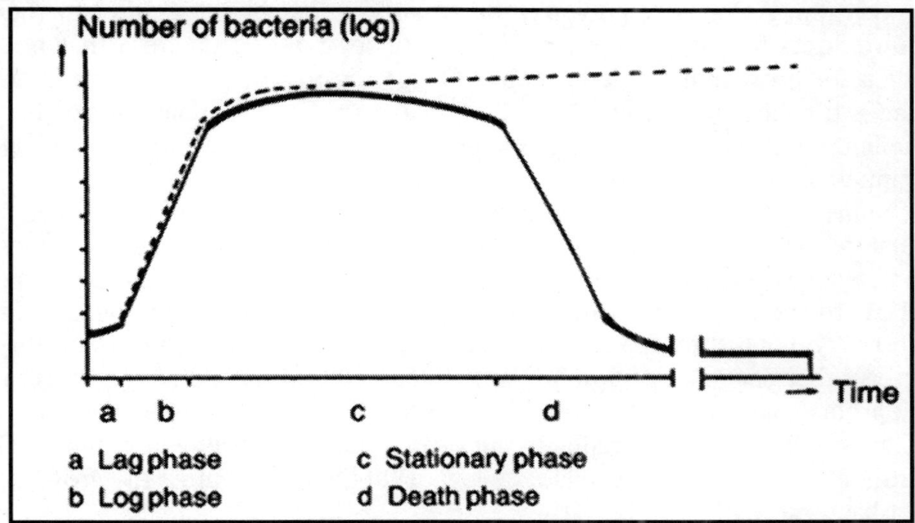

Fig. The Four Phases of Bacterial Growth.

Bacteria can be classified according to temperature preference: Psycrophilic bacteria grow at temperatures below 16°C, mesophilic bacteria grow best at temperatures between 16 and 40°C, and thermophilic bacteria grow best at temperatures above 40°C. Nutrients: Bacteria need nutrients for their growth and some need more nutrients than others. Lactobacilli live in milk and have lost their ability to synthesise many compounds, while Pseudomonas can synthesise nutrients from very basic ingredients.

Bacteria normally feed on organic matter; as well as material for cell formation organic matter also contains the necessary energy. Such matter must be soluble in water and of low molecular weight to be able to pass through the cell membrane. Bacteria therefore need water to transport nutrients into the cell.

If the nutrient material is not sufficiently broken down, the micro-organism can produce exo-enzymes which split the nutrients into smaller, simpler components so they can enter the cell. Inside the cell the nutrients are broken down further by other enzymes, releasing energy which is used by the cell.

Water: Bacteria cannot grow without water. Many bacteria are quickly killed by dry conditions whereas others can tolerate dry conditions for months;

bacterial spores can survive dry conditions for years. Water activity (AW) is used as an indicator of the availability of water for bacterial growth. Distilled water has an AW of 1. Addition of solute, e.g. salt, reduces the availability of water to the cell and the AW drops; at AW less than 0.8 cell growth is reduced. Cells that can grow at low AW are calledosmophiles.

Oxygen: Animals require oxygen to survive but bacteria differ in their requirements for, and in their ability to utilise, oxygen. Bacteria that need oxygen for growth are called aerobic. Oxygen is toxic to some bacteria and these are called anaerobic. Anaerobic organisms are responsible for both beneficial reactions, such as methane production in biogas plants, and spoilage in canned foods and cheeses.

Some bacteria can live either with or without oxygen and are known as faculative anaerobic bacteria. Acidity: The acidity of a nutrient substrate is most simply expressed as its pH value. Sensitivity to pH varies from one species of bacteria to another. The terms pH optimum and pH maximum are used. Most bacteria prefer a growth environment with a pH of about 7, i.e. neutrality. Bacteria that can tolerate low pH are called aciduric. Lactic acid bacteria in milk produce acid and continue to do so until the pH of the milk falls to below 4.6, at which point they gradually die off. In canning citrus fruits, mild heat treatments are sufficient because the low pH of the fruit inhibits the growth of most bacteria.

Bacteria in Milk

Milk fresh from a healthy cow contains few bacteria, but contamination during handling can rapidly increase bacterial numbers. Milk is an ideal food and many bacteria grow readily in it.Some bacteria are useful in milk processing, causing milk to sour naturally, leading to products such as irgo.

However, milk can also carry pathogenic bacteria, such asSalmonella, Tuberculosis bovis and Brucella, and can thus transmit disease. Other bacteria can cause spoilage of the milk, and spoilage and poor yields of products.

MOULDS

Moulds are a heterogeneous group of multicelled organisms which reproduce asexually either by spore formation or by fragmentation. They can grow on a wide variety of substrates and are generally regarded as spoilage organisms.

However, moulds are used in the production of antibiotics and in certain cheese varieties. Moulds are aerobic organisms and their growth on foods can be retarded by excluding air through careful packaging. They can be killed by relatively mild heat treatments, but mould spores are more resistant to heat.

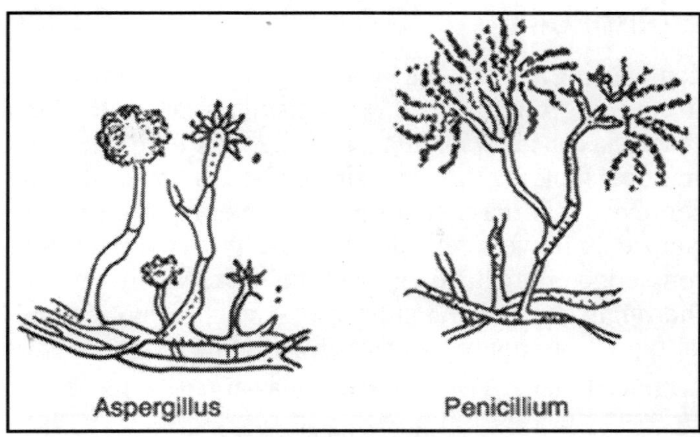

Fig. Structure of Moulds.

YEASTS

Yeasts are unicellular organisms which reproduce asexually by budding. They are used industrially to ferment carbohydrates to such products as alcohol and citric acid. Yeasts are not usually used in milk processing and are normally regarded as spoilage organisms in dairy products.

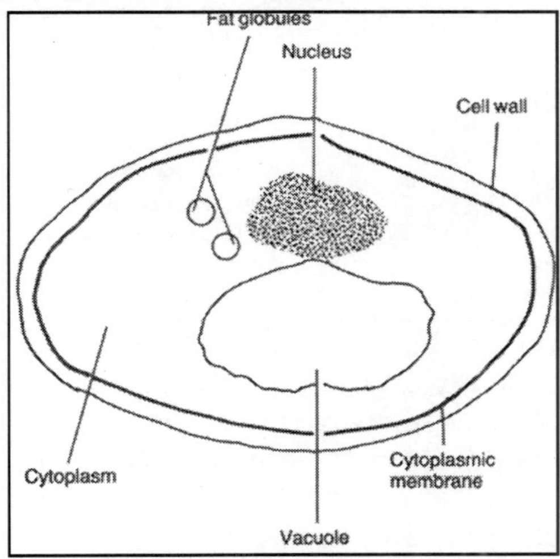

Fig. Structure of a Yeast Cell.

VIRUSES

Viruses are extremely small organisms comprising a spherical head containing the genetic material and a cylindrical tail. They cannot reproduce themselves, and must invade other cells in order to reproduce. Viruses that attack bacterial cells are known as bacteriophages: bacteriophages that attack acid-producing bacteria inhibit acid production in milk.

MILK MICROBIOLOGY

In addition to being a nutritious food for humans, milk provides a favourable environment for the growth of microorganisms. Yeasts, moulds and a broad spectrum of bacteria can grow in milk, particularly at temperatures above 16°C. Microbes can enter milk via the cow, air, feedstuffs, milk handling equipment and the milker. Once microorganisms get into the milk their numbers increase rapidly. It is more effective to exclude micro-organisms than to try to control microbial growth once they have entered the milk. Milking equipment should be washed thoroughly before and after use—rinsing is not enough.

Bacterial types commonly associated with milk.

Table. Bacterial Types Commonly Associated with Milk.

Pseudomonas	Spoilage
Brucella	Pathogenic
Enterobacteriaceae	Pathogenic and spoilage
Staphylococci	
Staphylococcus aureus	Pathogenic
Streptococcus	
S. agalactiae	Pathogenic
S. thermophilus	Acid fermentation
S. lactis	Acid fermentation
S. lactis-diacetyllatic	Flavour production
S. cremoris	Acid fermentation
Leuconostoc lactis	Acid fermentation
Bacillus cereus	Spoilage
Lactobacillus	
L. lactis	Acid production
L. bulgaricus	Acid production
L. acidophilus	Acid production
Propionibacterium	Acid production
Mycobacterium tuberculosis	Pathogenic

Microbial growth can be controlled by cooling the milk. Most microorganisms reproduce slowly in colder environments. Cooling milk also slows chemical deterioration. The temperature of freshly drawn milk is about 38°C. Bacteria multiply very rapidly in warm milk and milk sours rapidly if held at these temperatures. If the milk is not cooled and is stored in the shade at an average air temperature of 16°C, the temperature of the milk will only have fallen to 28°C after 3 hours. Cooling the milk with running water will reduce the temperature to 16°C after 1 hour. At this temperature bacterial growth will be reduced and enzyme activity retarded. Thus, milk will keep longer if cooled.

Natural souring of milk may be advantageous: for example, in smallholder butter-making, the acid developed assists in the extraction of fat during churning. The low pH retards growth of lipolytic and proteolytic bacteria and therefore protects the fat and protein in the milk. The acidity of the milk also inhibits the growth of pathogens. It does not, however, retard the growth of moulds. Naturally soured milk is used to make many products, e.g. irgo, yoghurt, sour cream, ripened buttermilk and cheese. These products provide ways of preserving milk and are also pleasant to consume. They are produced by the action of fermentative bacteria on lactose and are more readily digested than fresh milk.

The initial microflora of raw milk reflects directly microbial contamination during production. The microflora in milk when it leaves the farm is determined by the temperature to which it has been cooled and the temperature at which it has been stored. The initial bacterial count of milk may range from less than 1000 cells/ml to 106/ml. High counts (more than 105/ml) are evidence of poor production hygiene. Rapid tests are available for estimating the bacterial quality of milk.

MILK PASTEURISATION

Pasteurisation is the most common process used to destroy bacteria in milk. In pasteurisation, the milk is heated to a temperature sufficient to kill pathogenic bacteria, but well below its boiling point. This also kills many non-pathogenic organisms and thereby extends the storage stability of the milk. Numerous time/temperature combinations are recommended but the most usual is 72°C for 15 seconds followed by rapid cooling to below 10°C. This is normally referred to as High Temperature Short Time (HTST) treatment. It is carried out as a continuous process using a plate heat-exchanger to heat the milk and a holding section to ensure that the milk is completely pasteurised. Milk is normally pasteurised prior to sale as liquid milk. Pasteurisation is used to reduce the microbial counts in milk for cheese-making, and cream is pasteurised prior to tempering for butter-making in some factories.

Batch pasteurisation is used where milk quantities are too small to justify the use of a plate heat-exchanger. In batch pasteurisation, fixed quantities of milk are heated to 63°C and held at this temperature for 30 minutes. The milk is then cooled to 5°C and packed. The lower temperature used for batch pasteurisation means that a longer time is required to complete the process—30 minutes at 63°C, compared with 15 seconds a 72°C.

Effects of Pasteurisation on Milk

Pasteurisation reduces the cream layer, since some of the fat globule membrane constituents are denatured. This inhibits clustering of the fat globules and consequently reduces the extent of creaming. However, pasteurisation does not reduce the fat content of milk. Pasteurisation has little effect on the nutritive value of milk. The major nutrients are not altered. There is some loss of vitamin C and B group vitamins, but this is insignificant. The process kills many fermentative organisms as well as pathogens. Micro-organisms that survive pasteurisation are putrefactive. Although pasteurised milk has a storage stability of 2 to 3 days, subsequent deterioration is cause by putrefactive organisms. Thus, pasteurised milk will putrefy rather than develop acidity.In rural milk processing, many processes depend on the development of acidity, and hence pasteurisation may not be appropriate.

MILK STERILISATION

In pasteurisation, milk receives mild heat treatment to reduce the number of bacteria present. In sterilisation, milk is subjected to severe heat treatment that ensures almost complete destruction of the microbial population. The product is then said to be commercially sterile. Time/ temperature treatments of above 100°C for 15 to 40 minutes are used. The product has a longer shelf life than pasteurised milk. Another method of sterilisation is ultra-heat treatment, or UHT. In this system, milk is heated under pressure to about 140°C for 4 seconds. The product is virtually sterile. However, it retains more of the properties of fresh milk than conventionally sterilised milk.

MICROBIOLOGY OF BUTTER

Butter is made as a means of extracting and preserving milk fat. It can be made directly from milk or by separation of milk and subsequent churning of the cream.

Sources of Contamination

In addition to bacteria present in the milk other sources of bacteria in butter are (1) equipment, (2) wash water, (3) air contamination, (4) packing materials, and (5) personnel.

Equipment

In smallholder butter-making, bacterial contamination can come from unclean surfaces, the butter maker and wash water. Packaging materials, cups and leaves are also sources of contaminants. Washing and smoking the churn reduces bacterial numbers. But traditional equipment is often porous and is therefore a reservoir for many organisms.

When butter is made on a larger processing scale, bacterial contamination can come from holding-tank surfaces, the churn and butter-handling equipment. A wooden churn can be a source of serious bacterial, yeast and mould contamination since these organisms can penetrate the wood, where they can be destroyed only by extreme heat. If a wooden churn has loose bands, cream can enter the crevices between the staves, where it provides a growth medium for bacteria which contaminate subsequent batches of butter. However, if care is taken in cleaning a wooden churn this source of contamination can be controlled. Similar care is required with scotch hands and butter-working equipment.

Wash water

Wash water can be a source of contamination with both coliform bacteria and bacteria associated with defects in butter. Polluted water supplies can also be a source of pathogens.

Air

Contamination from the air can introduce spoilage organisms: mould spores, bacteria and yeasts can fall on the butter if it is left exposed to the air. Moulds grow rapidly on butter exposed to air.

Packaging

Care is required in the storage and preparation of packaging material. Careless handling of packaging material can be a source of mould contamination.

Personnel

A high standard of personal hygiene is required from people engaged in butter-making. For example, in New Zealand the 1938 dairy produce regulations stated "no person shall permit his bare hands to be brought in contact with any butter at any time immediately following manufacture or during the wrapping, packaging, storage and transport of such butter". Personnel pass organisms to butter via the hands, mouth, nasal passage and clothing. Suitable arrangements for disinfecting hands should be provided, and clean working garments should not have contact with other clothes.

CONTROL OF MICRO-ORGANISMS IN BUTTER

Salting effectively controls bacterial growth in butter. The salt must be evenly dispersed and worked in well. Salt concentration of 2% adequately dispersed in butter of 16% moisture will result in a 12.5% salt solution throughout the water-in-oil emulsion.

Washing butter does little to reduce microbiological counts. It may be desirable not to wash butter, since washing reduces yield. The acid pH of serum in butter made from ripened cream or sour milk may control the growth of acid-sensitive organisms.

Microbiological analysis of butter usually includes some of the following tests: total bacterial count, yeasts and moulds, coliform estimation and estimation of lipolytic bacteria. Yeast, mould and coliform estimations are useful for evaluating sanitary practices.

The presence of defect producing types can be indicated by estimating the presence of lipolytic organisms. All butter contains some micro-organisms. However, proper control at every stage of the process can minimise the harmful effects of these organisms.

TYPES OF MICROORGANISMS AND THEIR ACTIVITY IN MILK

The numbered list below identifies seven types of bacteria according to how they change the properties of milk. Often these changes are negative (spoilage) but as we will see in later sections, many of these bacteria are important to the development of cheese flavour. Before proceeding to the list, please note the following definitions:

- Psychrotrophic refers to microorganisms which are able to grow at temperatures less than 7C. Cold milk storage and transport selects for psychrotrophic bacteria which are often proteolytic and lipolytic. Common psychrotrophic bacteria in milk are species of Micrococci, Bacilli, Staphyloccoci, Lactobacilli, Pseudomonas, and coliforms. Pseudomonas species are the most common and typically have the most impact on quality. At temperatures of 2 - 4C, bacterial growth in milk is mainly due to strains of Pseudomonas flourescens. Little growth occurs at temperature less than 2C.
- Spore forming bacteria are able to exist in a highly stable form called 'spores'. In the spore state, these bacteria are able to withstand greater extremes of acidity, temperature and dessication.
- Enzymes are biological catalysts that accelerate the rates of biochemical reactions. Bacterial enzymes are most significant to milk spoilage and cheese ripening but it is important to distinguish between the enzyme and the bacterial source. For example, many psychrotrophic bacteria produce heat stable enzymes which remain active in milk and cheese even after the bacteria are killed by pasteurization.

Keeping the above definitions in mind, note the following types of microorganisms, grouped according to their impact on milk quality.

- Lactic acid bacteria which ferment lactose to lactic acid and other end products. Lactic acid bacteria (LAB) important to cheese making will be described further in Chacultures7. For now note the following:
  - As noted earlier, LAB are able to readily metabolize lactose so they have some competitive advantage over other microorganisms.
  - Notwithstanding, their ability to metabolize lactose, LAB prefer temperatures greater than 30C, so, depending on initial relative counts, psychrotrophic bacteria including some coliform and pseudomonas bacteria are able to outgrow LAB at room temperature.
  - Proteolytic bacteria which degrade protein and cause bitterness and putrefaction. Most important in cheese milk are species of:

  - Pseudomonas which are psychrotrophic and produce heat stable lipases.
  - Bacillus which form heat stable spores and survive pasteurization
- Lipolytic bacteria which degrade fats and produce lipolytic rancidity. Again, the most common example in milk is the genus Pseudomonas. Several psychrotrophic species of Pseudomonas produce heat stable lipases as well as proteases.
- Gas producing microorganisms which cause cheese openness, floating curd in cottage cheese, and gassy milk.
  - Yeasts are always present in milk and are common contaminants during the cheese making process. They may cause 'yeast slits' in cheese and contribute to ripening of surface ripened cheese.
  - Coliform bacteria are always present in milk but their numbers can be minimized by good sanitation. Also, coliform bacteria compete poorly with lactic acid bacteria, so their numbers rapidly decrease in the presence of a rapidly growing lactic acid culture.
  - Clostridium tyrobutyricum is a thermoduric (survives pasteurization) spore forming organism of legendary fame among cheese makers. C. tyrobutyricum causes gas formation (carbon dioxide) during the later stages of ripening of Swiss and Dutch type cheeses. The resulting craters and cracks in the cheese are called 'late gas defect'. European cheese makers frequently check raw milk for thermoduric and/or spore forming bacteria to estimate potential for late gas defects. Five hundred spores per litre of milk are sufficient to cause late gas defect.
  - Propioni bacterium produces the desirable gas formation in Swiss type cheese.
  - Some lactic cultures, called heterofermentative, also produce carbon dioxide. See Cultures.
- Ropy bacteria cause stringy milk due to excretion of gummy polysaccharides. Usually ropy bacteria such as Alcaligenes viscolactis are undesirable. However, in some fermented dairy products, ropy lactic acid bacteria such as certain subspecies of Lactococcus lactis are used to develop texture.
- Sweet curdling bacteria produce rennet-like enzymes which may coagulate milk. Common examples are the psychrotrophic spore formers Bacillus subtilis and Bacillus cereus.
- Numerous off flavours have been associated with specific milk contaminates.

MICRO-ORGANISMS IN MILK

Milk is sterile at secretion in the udder but is contaminated by bacteria even before it leaves the udder. Except in the case of mastisis, the bacteria at this point are harmless and few in number. Further infection of the milk by microorganisms can take place during milking, handling, storage, and other pre-processing activities.

Minute living cells which can not be seen by the naked eyes and are visible only under microscope are called micro- organisms or microbes and the science which deals with them is known as Microbiology (Micros means small and Bios means life. This term was coined by 'Louis Pastcur'.

Micro-organisms are present every where in nature, in soil, in water, in air, in all climatic conditions in the extremes of the polar region and the tropics, in hot dry deserts and in the wet marshy lands. They arc found in the high altitudes of the atmosphere and even in the deep wells of the petroleum fields and in coal mines. They are found even inside the intestines of animals and men.

The micro-organisms which arc most important in dairy are the bactcria, yeasts and moulds. All of these are members of plant kingdom, that are characterised as plants without chlorophyll, the green colouring matter of higher plants. The dairy industry is vitally concerned about these micro-organisms because of their ability to produce several changes in milk and milk products.

Milk as it exists in the udder of healthy animal is considered to be free from micro-organism. As it leaves the udder various micro-organisms gain entry into the milk. Their growth is faster when the conditions for their growth rate is favourable. Since, milk contains all the nutrients required for the growth of micro- organisms and serves as a good medium for their growth, their number increases in geometrical progression. If the growth of these micro-organisms is not checked, they continue to grow and multiply and produce several bio-chemical changes which are considered undesirable in milk.

CLASSIFICATION

Thc micro-organisms can be classified into two broad categories: pathogenic and non pathogenic.

Pathogenic Orgamisms

The micro-organisms which are capable of producing disease are called pathogenic organisms. If these micro-organisms are present in milk or milk products and are not destroyed before consumption then they are likely lo cause disease in the body. Their presence in milk as well as in milk products is highly objectionable. For example dysentery causing microbes Shigella dysenteriae.

Non-pathogenic Organisms

Non-pathogenic micro-organisms do not cause disease in the body. Their presence is also undesirable in milk because they are capable causing spoilage, although several non-pathogenic organisms have been found useful in the manufacture of various milk products. For example, species streptococcus lactis sours milk. Some of the important non-pathogenic bacteria which can be used for are mentioned here.

Yeasts

The yeasts most frequently encountered in milk and milk products act upon the lactose to produce acid and carbon dioxide. These micro-organisms are used in the preparation of a few special milk products. Yeasts are most numerous in the raw cream of market during hot weather, but they are potential contaminants throughout the year.

Moulds

Whereas bacteria and yeasts are basically uni-cellular, moulds are multi-cellular organisms. This means several cells constitute a mould plant. Some of these cells perform special functions for the whole plant. Moulds are sometimes observed on the surfaces of butter, old cream, khoa or cheese. They often grow in large concentrations which are visible as a fuzzy or fluffy growth. Moulds are black, blue, green, grey, or white in colour. Apart from discolouring milk products, moulds often produce desirable flavours and odours. In a few cases, such as certain kinds of cheese, moulds arc essential. The premises and products of dairy processing industry must be protected from mould contamination.

Bacteria

Bacteria are the most common and probably the most numerous of the micro-organisms with which the dairy processing industry is concerned. They seem to be present everywhere. Some of the important types of bacteria which are concerned with dairy industry are mentioned here.

- Lactose fermenting bacteria: Lactose fermenting group of bacteria act on lactose in milk and convert it into lactic acid. The examples of this group of bacteria are Streptococcus lactis, Streptococcus cremolis, Lactobacillus thermophilus, Lactobacillus acidophilus and Escherichia coli.

These bacteria can further be classified under the following sub-groups.

Homo-fermentative: Those micro-organisms which act on lactose and produce only lactic acid as a principal product and very small quantity of carbondi-oxide, are known as homofermcntative bacteria.

Lactose + Homofermcntative Bactcria > Lactic acid

Hetero-fermentative: Those organisms which produce certain other substances apart from lactic acid from lactose are known as hctcrofermentalive.

Lactose + heterofcrmentative > Lactic acid + Acetic acid Bacteria acid bacteria + CO_2 (50%)

Proteolytic bacteria: Proteolytic bacteria act on milk proteins and hydrolyse them. The hydrolysis of protein is accompanied by an increase in pH. This is brought about by the enzyme renin produced by the proteolytic bacteria when coagulation of milk without the presence of appreciable quantity of lactic acid occurs. It is termed as sweet curdling. Sweet curdling is normally followed by some degree of protein digestion. Proteolysis in milk is characterised by alkaline reaction and bitter taste. Some example of proteolytic bacteria are Bacillus subtilis, Bacillus cereus, Bacillus calidolactis, Pseudomonas Putrificiens, etc.

Lipolytic bacteria: The bacteria which split fats are called lipolytic bacteria. They produce enzymes (Iipases) which hydrolyse milk fat to glycerol and fatty acids. Some of these fatly acids have a sharp flavour and odour and are responsible for imparting rancid flavour in milk. Some of the examples of lipolytic bactcria are Pseudomonas fragi, P.fluoroscents, Acllromobacter lipolyticum ctc.

EFFECTS OF BACTERIA

Ropiness and frothiness are the most common effect of micro- organisms.

Lactic Acid Bacteria

This group of bacteria are able to ferment lactose to lactic acid. They are normally present in the milk and are also used as starter cultures in the production of cultured dairy products such as yogurt. Note: many lactic acid bacteria have recently been reclassified; the older names will appear in brackets as you will still find the older names used for convenience sake in a lot of literature. Some examples in milk are:

- Lactococci
  - L. delbrueckii subsp. lactis (Streptococcus lactis)
  - Lactococcus lactis subsp. cremoris (Streptococcus cremoris)
- Lactobacilli
  - Lactobacillus casei
  - L.delbrueckii subsp. lactis (L. lactis)
  - L. delbrueckii subsp. bulgaricus

(Lactobacillus bulgaricus)

- Leuconostoc

Coliforms:

coliforms are facultative anaerobes with an optimum growth at 37° C. Coliforms are indicator organisms; they are closely associated with the presence of pathogens but not necessarily pathogenic themselves. They also can cause rapid spoilage of milk because they are able to ferment lactose with the production of acid and gas, and are able to degrade milk proteins.

They are killed by HTST treatment, therefore, their presence after treatment is indicative of contamination.Escherichia coli is an example belonging to this group.

Significance of microorganisms in milk:

- Information on the microbial content of milk can be used to judge its sanitary quality and the conditions of production
- If permitted to multiply, bacteria in milk can cause spoilage of the product
- Milk is potentially susceptible to contamination with pathogenic microorganisms. Precautions must be taken to minimize this possibility and to destroy pathogens that may gain entrance
- Certain microorganisms produce chemical changes that are desirable in the production of dairy products such as cheese, yogurt.

SPOILAGE MICROORGANISMS IN MILK

The microbial quality of raw milk is crucial for the production of quality dairy foods. Spoilage is a term used to describe the deterioration of a foods' texture, colour, odour or flavour to the point where it is unappetizing or unsuitable for human consumption. Microbial spoilage of food often involves the degradation of protein, carbohydrates, and fats by the microorganisms or their enzymes.

In milk, the microorganisms that are principally involved in spoilage are psychrotrophic organisms. Most psychrotrophs are destroyed by pasteurization temperatures, however, some like Pseudomonas fluorescens, Pseudomonas fragi can produce proteolytic and lipolytic extracellular enzymes which are heat stable and capable of causing spoilage. Some species and strains of Bacillus, Clostridium, Cornebacterium, Arthrobacter, Lactobacillus, Microbacterium, Micrococcus, and Streptococcus can survive pasteurization and grow at refrigeration temperatures which can cause spoilage problems.

Pathogenic Microorganisms in Milk

Hygienic milk production practices, proper handling and storage of milk, and mandatory pasteurization has decreased the threat of milkborne diseases such as tuberculosis, brucellosis, and typhoid fever. There have been a number of foodborne illnesses resulting from the ingestion of raw milk, or dairy products made with milk that was not properly pasteurized or was poorly handled causing post-processing contamination. The following bacterial pathogens are still of concern today in raw milk and other dairy products:

- Bacillus cereus
- Listeria monocytogenes
- Yersinia enterocolitica

- Salmonella spp.
- Escherichia coli O157:H7
- Campylobacter jejuni

It should also be noted that moulds, mainly of species of Aspergillus, Fusarium, and Penicillium can grow in milk and dairy products. If the conditions permit, these moulds may produce mycotoxins which can be a health hazard.

Ropiness

Some bacteria produce sticky or gum like substance which cause ropiness in milk. This condition becomes apparent when milk forms long threads of several inches or even feet in length when poured from a container. The common organisms responsible for ropiness in milk are Alcaligenes viscosus and Aerobacter aerogenes.

Frothiness

Formation of foam in milk as well as in cream may be due to action of micro-organisms. Certain organisms produce carbon di-oxide which causes foaming and this is normally accompanied by off-flavour in milk as well as in cream. The most common bacteria which cause frothiness in milk is Aerobacter aerogenes. Sometimes yeast may also cause frothiness and the common examples are Tonlla cremoris and Tonda spherica.

FACTORS AFFECTING THE GROWTH OF MICRO-ORGANISMS

The growth of micro-organisms is influenced by the following factors: (A) food, (B) moisture, (C) oxygen, (D) chemical and physical environment, (E) preservatives, (F) light, (G) concentration of additives, and (H) temperature.

Food

The food requirements of micro-organism are extremely small quantitatively, but it depends on the quality some of the elements like carbon, oxygen, hydrogen, nitrogen, phosphorous, and sulphur are essential. They may be obtained from proteins, fats and carbohydrates present in milk. Mineral elements are also necessary in minute quantities for the development of the cells, especially potassium, calcium, sodium; magnesium, manganese, iron and chlorine.

Moisture

Water is as absolute requirement for the growth of micro- organisms. Micro-organisms find the optimum condition for growth when the food stuff and atmosphere are saturated with moisture.

Oxygen

Micro-organisms are specific as far as atmosphere is concerned. Certain organisms require unlimited supply of oxygen for their growth while some do not grow if traces of free oxygen is present. Between these two extremes the majority of micro-organisms survive under varying concentration of oxygen. Those organisms which require oxygen for their growth are, called aerobic bacteria while anaerobic bacteria do not require oxygen for their growth.

Chemical and Physical Environment

Certain micro-organisms are sensitive to the presence of strong acids and bases. Though most of them prefer a medium which is nearly neutral in reaction, yet they can tolerate certain amount of acidity as well as alkalinity. The growth of most of the acid producing organisms is retarded when a considerable amount of acid is produced by them in the medium. The growth of most of the acid producing organism is retarded when a considerable amount of acid is produced by them in the medium. Yeasts and moulds usually grow best in acid media and often initiate their growth after the acidity has become concentrated sufficiently to check bacterial growth.

Preservatives

These substances check the growth of micro-organisms and improve the keeping quality of the substance in which they are present. These are known as preservatives. Formalin, boric acid, borax, benzoic acid, salicylic acid, mercuric chloride, hydrogen peroxide and other chemicals are used in milk as preservatives. The purpose of using preservative is to prolong its keeping quality, although addition of any preservative in milk intended for human consumption is illegal and criminal.

Light

The majority of the micro-organisms grow best in the dark. Light is generally harmful to their growth. Bright light whether natural or artificial has a tendency to check the growth or destroy them. Direct sunlight and ultra-violet rays are used for destroying the micro-organisms. Since, these sources of light affect the quality of milk, they arc not used for destroying the micro-organisms in milk.

Concentration of Additives

The growth of micro-organisms is encouraged when the concentration of additive such as salt or sugar is low. In case these substances are present in concentrated forms, then the growth of micro-organisms is retarded. For example, when sugar or salt concentration in some substances are added in adequate amounts the keeping quality of the product increases.

Temperature

Temperature is one of the most important factor which affects the growth of micro-organisms. On the basis of temperature bacteria can be classified into three categories, which are mentioned here.

1. *Psychrophilic bacteria:* Psychrophylic bacteria are cold loving bacteria. They prefer lower temperature for their growth. They grow within the temperature range of 3 to 20°C, where as optimum temperature for their growth is about 7° Centigrades. Some of the examples of this type of bacteria are Pseudomonas fragi, Pseudomonas fluorescensetc.
2. *Mesophilic bacteria:* Mesophilic bacteria can tolerate relatively low as well as higher temperature for their growth. They have a range of20 -50 °C for their growth. The optimum temperature range for this group of micro-organisms is about 37° Centigrades. The majority of the organisms in atmosphere is mesophilic. Some of the bacteria of this group are Streptococcus lactis, Streptococcus cremoris etc.
3. *Thermoduric bacteria:* Thermoduric bacteria grow best at a temperature of about 20 -36° Centigrades. They have a peculiarity that neither they are destroyed nor they grow at a higher temperature and because of this character they are important from the pasteurization point of view. Some examples of thermoduric bacteria are Streptococcus thernlophihlS, Streptococcus duran, Micrococcus hlteus, Microbacterium flavum etc.
4. *Thermophilic bacteria:* Thermophilic bacteria are heat loving. They grow at a higher temperatures.The optimum temperature for their growth is about 50-700 centigrades. Some examples of thermophilic bacteria are Lactobacillus thermophilus, Bacillus calidoactics etc.

MICRO-ORGANISMS ACTION

The micro-organisms act in the milk and produce or change oudors, coloures flavours, gas enzymes and alkali. There are briefly explained here.

Odours

An acid odour can often be detected in milk having slight developed acidity. Micro-organisms acting upon both the proteins and the fat oflen produce changes which are detected by odour. Milk is frequently graded on the basis of its odours.

Colour

Moulds often discolour the surface of the butter and cheese, and may carry their discolouraion into the interior of these products. In the absence of proper sanitation, Pseudomonas spp may give rise to a yellowish film on product surfaces contact of equipment.

Flavour

The acid flavour commonly found in milk is caused by micro-organisms which attack the lactose. A bitter flavour results from certain types of protein breakdown caused by micro-organisms. Mould contamination in milk products, mainly cream, butter, cheece and khoa may give undesired mouldy or nasty flavours. The quality of flavour in dahi, olher fermented milk, butter and cheesc is improved by the use of sele-cted micro-orgamisms.

Gas

Many micro-organisms, usually reaching with the lactose, produce gas in milk and milk products. Yeasts are particularly prone to gas, mainly carbon di-oxide. Gas formation is a common defect of sour cream. Gas production is essential in certain types of cheese such as, Emmanthal, and in one or two fermented milks such as kumiss.

Enzymes

Some enzymes are produced in milk and milk products, but few of the natural enzymes of milk are attacked or changed by the action of micro-organisms.

Alkali

A few micro-organisms such as Alcaligenes viscolactis, Pseudomonas fluorescens, and Micrococcus ureae, produce alkali or neuteralise acidity developed in milk. Yeasts and moulds are seldom responsible for this action in milk.

MICRO-ORGANISMS IN PREPARATION OF MILK PRODUCTS

The following are the major milk product in the preparation of which micro-organism are involved.

Fermented Milk

Curd or dahi is a one of the fermented milk product. The species of lactic acid bacteria occuring most commonly in dahi include Lactobacillus bulgaricus, Streptococcus thermophilis, S. faecalis, S. lactis, L. casei and L. plantarum. The term “yoghurt” is widely used in other milk producing countries for products similar to duhi.

Streptococcus themlophillis and Lactobacililis bulgaricus are used together for making yoghurt, the former produces some acid and a fine aroma, the latter one produces high acidity. The wide variety of fermented milks includes not only those coagulated by acid formation, but also are characterised by flavours produced by bacteria and yeasts.

Cheese

The microtlora of cheese may differ greatly from one kind of cheese to another. The starter or inoculum used must vary accordingly. Here some of the varieties of bacteria and moulds employed in cheese making are mentioned. In cheddar cheese, the action of rennet is influenced by acidity. The starter used for this cheese, therefore, must contain acid producing bacteria. The most commonly used in this starter arc Streptococcus lactis or S. cremoris.

Roquefort or Blue cheese is made with the help of lactic acid starter and rennet, the bacteria is usually S. lactis. As the curd is dipped, spores of Penicillium roquefort (blue green mould) spreads throughout the curd. After being pressed the cheese is punctured to admit air to facilitate mould growth in its interior. Camembert cheese is similar to Roquefort. It is prepared in smaller prints and cured with P. camemberti which is a white mould.

Ghee

Ghee or clarified butter is extensively used in Indian homes. The micro biological problems encountered in its preservation are very few as it contains only a very little amount of moisture (less than 1%). It is almost pure butter fat and so is resistant against the attack of bacteria, except a few species of lipolytic organisms. If the Ghee is not being packed under hygienic environment, it shows that it contains a large amount of bacteria. It contains micrococci, gram negative rods, spore forming mesophilic rods and moulds of the types of Aspergillus and Penicillium spp. The keeping quality of ghee depends primarily upon its moisture content.

Concentrated Milk Products

The two most common concentrated milk products found in many parts of India are khoa and rabri.

- *Khoa:* The keeping quality of khoa depends entirely upon the degree of contamination and the extent to which it is exposed to conditions favourable to the growth of micro-organisms subsequent to its manufacture. During the process of manufacture of khoa continued boiling process is followed. Almost all the bacteria present in milk arc killed, still a few spore formers may survive. Therefore, the bacteria present in khoa are mostly due to the faulty methods of its handling after preparation.
- *Rabri:* Rabri is another product of milk that is prepared by heating in a wide top open pan. In this case the milk is not boiled, but is heated to a temperature at which a skin may form on the surface. The skin which forms during heating is carefully removed from time to time by the use of a special dipper, usually made of bamboo. During the process of making, equipments used arc not sterilized and the bamboo

dipper may act as a source of contamination. Besides these sources handling it after preparation also may result in contamination of the product. The most common organisms found in rabri are spore forming rods, micrococci, moulds and yeasts. Some disease producing bacteria may also enter into it.

Chhenna

Chhenna is one of the indigenous milk product in India; It is prepared from milk by precipitation of the protein, by the addition of an acid, either sour whey, citrus fruit juice or citric acid solution. The chhenna is then removed by draining off the whey. Chhenna also has a very low keeping quality and it tends to become sour rather rapidly. It is highly perishable and its keeping quality is only 6 -8 hours in summer and 16-20 hours in winter. Though it is prepared after the milk has been boiled, spore forming aerobic rods survive boiling and they get the most favourable conditions for growth and at summer temperature they start growing very rapidly.

The types of bacteria present in chhenna represent micrococci, aerobic spore formers and non-spore forming rods. Mould growth on the surface is visible within 48 hours. The common moulds found in chhenna belong to the species of Penicillium aspergillus, Mucos, Rhizopus, Fusarium and Paecelomyces. Most of the bacterial species are thermoduric and the spore formers are all proteolytic. Moulds are also very active proteolytic agents.

Hygienically prepared chhenna may keep well for a longer period. It may be preserved by adding 0.5% sodium benzoate, or 0.15 to 0.2% sodium propionate. In both the cases the keeping quality is increased up to 5 to 6 days. Sugar may also be added as a preservative. Thirty percent of sugar has been found to be more effective and it keeps well for 4 to 5 days. The wrapping materials also have great influence on the keeping quality of chhenna. Mostly butter paper should be used for wrapping. It has also been noted that if the paper is rubbed with 0.5% solution of sodium benzoate or propionate the keeping quality is increased to a great extent.

Ice-Cream

Bacterial content of ice-cream is dependent on the quality of, the ingredients used. The ingredients used in ice-cream may include cream with or without added milk solids, sugar, gelatin, stabilizer, and eggs or egg solids. Once the ingredients are frozen micro-organisms cannot multiply in it, Number of bacteria: Cream content in ice-cream varies widely in bacteriological condition. Pasteurization of the cream normally destroys a high percentage of the bacteria. Thcrmoduric bactcria, such as Micrococci and Stereptococci may be present in the cream. Utensils used to hold milk ice-cream mix should be thoroughly sterilized before using.

Egg products and gelatin may contribute large number of bacteria, including types of ice-creams. Sugar, in general, contains only small number of bacteria and is of little importance as a source of organisms in ice-cream. Flavouring and colouring materials, fruits and nuts may constitute important sources of bacteria since, they often are added in the raw state just as the mix enters the freezer. These nuts carry a wide variety of organisms including aerobic Sporformers, Micrococci and E.coli.

Types of bacteria: A wide variety of bacterial species have been isolated from ice-cream. These include Micrococci, Streptococci, aerobic and anaerobic spore formers, E. coli, A. aerogenes, Pseudomonas spp. and many other pathogenic and non-pathogenic bacteria. Acid forming species make up a considerable percentage of the flora. If favourable conditions for growth of bacteria are provided a variety of organisms grow in ice-creams and these include yeasts and moulds.

Disease transmission: Commercial ice-cream is known to cause many outbreaks of epidemics; Home-made ice-cream prepared with raw ingredients also may cause outbreaks of certain epidemics. Most of the outbreaks are due to food poisoning. Food poisoning is usually due to the growth of micrococcus Pyogenes aureus. It produces toxins 'that cause nausea, dizziness, headaches and diarrhoea.

Pasteurization of ice-cream mix destroys the Micrococci, but the toxins are not destroyed. Care should therefore, be taken to avoid contamination of the mix at high temperature, infection from the ice-cream makers and contamination from the utensils are some of the factors that are responsible for the outbreaks of diseases through ice-cream. A significant fact to be remembered is that if any disease-producing bacteria gets entry into the ice-cream it may be kept viable for years together in the frozen state.

Milk powder

Dried milk products represent the most stable products from the stand point of microbial spoilage. As long as they are kept in a dry state no micro-organisms of any type can develop in them. In fact the micro-organisms gradually decrease in numbers in dried foods during their storage. At present, milk is dried by the roller process and also by the various spray systems.

Since, the low content of moisture (less than 5%) absolutely prevents bacterial growth. In any storage temperature, the bacterial content of dried milk reflect the original position of milk and the conditions under which they were manufactured. Though the micro-organisms present In dried milk powder cannot multiply, they cause serious defects in the manufacture of different products. Bacillus mesentericus if present in the milk powder may cause a very serious defect known as ropy-bread originating from the use of such milk powder. When starter cultures arc prepared from dried milk, spore 10rmers 'present in it may give rise to serious troubles. Micro-organisms found in spray dried milk include thermoduric Streptococci such as S. themlophilus, micrococci, Lactobacilli, Achromobacter, aerobic and anaerobic spore forming types, moulds, yeast, coliform bacteria, and species of tiny rods known as microbacteria.

Butter

The quality of butter and its period of preservation depends greatly on the quality of the raw materials used in its preparation. The cream should be prepared from fresh clean milk produced and handled under hygienic conditions. Cream used for butter making is usually pasteurized at a temperature not below 85°C and therefore almost all the bacteria are destroyed except the spore formers.

Micro-organisms may gain entrance to cream after it has been pasteurized through the utensils and these organisms may include lactic acid bacteria, spore formers bacteria may gain entrance into butter through salt. Butter colour may act as a source of fungal mycelium in butter. However, butter colour and salt are not very great sources of bacteria in butter. Microflora of butter starter must contain S.lactis, S. cremoris, and the aroma producing diacetyl bacteria. In a starter the presence of S. lactis, S. cremoris, and S. diacetilactis is rnost favourable.

Colour Defects

Most of the colour defects in butter are caused by the microbial actions. Moulds that enter the butter from equipments, wrappers, or printers or from the air of the creamery or storage room. They produce brown, greenish, and black growth that discolours the surface of butter in addition to affecting the flavour.

Pseudomonas nigrifaciens enter butter from water and contaminate equipment. They produce a black pigment on the surface which may sometimes by mistaken for a grease smudge. Pink yeasts occasionally develop on the surface of butter and produce pink spots where they grow. Sulphur dioxide which sometimes escapes from the refrigerator may also cause pink discolouration on the surface of butter.

Moulds are generally present in the air and therefore it is very easy that butter gets contaminated by moulds. Cladosporium is generally found in abundance inside the butter and it is seen in dark patches. Mould growth in butter is an indication of the high moisture content in it. The higher the acidity of the butter more favourable is the conditions for the growth of moulds.

SOURCES OF MICRO-ORGANISMS IN MILK

The micro-organisms which infect the products come from different sources. Some of them are described in detail here.

Milch Animal

Unless the milk producing animal is clean and her flanks, udder and teats if not given special sanitary care just before milking, can be a source of considerable contamination. Dirt and hair often fall from her body into the milking utensils. The interior of the teats always contains micro-organism. The interior of the teats of the udder is warm, contains the last remnants of the milk flowing through it and hence serves the most ideal place for the growth of bacteria.

The first few streams of milk from each teat should be collected separately and discarded. The majority of the bacterial population of the interior of the udder is the micrococci. They are inert and produce no appreciable changes in milk. Besides the micrococci the udder microflora contains streptococci. Milk which contains only the microflora of the interior of the udder is called asceptic, milk. This type of milk might be obtained by taking strict asceptic precautions during milking.

Mastitis is a disease of the mammary tissues. It may be acute or chronic. Acute mastitis results from staphylococcus tubercular, or other infection with, agalactiae. Many pathogens like those associated with mastitis, may survive for considerable time in milk and milk products. Mastitis milk must be considered potentially unsafe.

Milkers Hands

In a successful dairying the attendants must be trained to follow the strict sanitation. It is through the unhygienic conditions that the pathogenic organisms may gain entrance to milk. It is desirable that the milkers should have clean and tight- fitting clothes. Their hands should be properly washed with soap and water before milking, the nails should be properly and regularly trimmed. Coughing and sneezing, even talking will spray droplets of nasal and oral discharges into the atmosphere.

Table. Amount of Bacteria in Cow Milk Under Different Conditions.

Sl.No	Conditions of production	Amt of bacteria in 1 ml milk			
		1st teat	2nd teat	3rd teat	4th teat
1	Cows kept without any special care for four days	620	1650	7004	500
2	Two days kept with special care and the opening of the teats were kept covered by collodion	22	12	15	3
3	Cows kept with special care for one day	120	250	63	180

Some of the serious diseases like septic sore throat, tuberculosis, etc. are spread through the milk supply, where the milkers have such unhygienic habits. A surgical mask is an effective addition to the uniform workers processing dairy products. The milkers should be examined regularly so that they are free from contagious diseases. The effect of asceptic precautions on the quantity of bacteria in milk. The bacterial quality of milk is mostly dependent on the care of the cows. If utmost care is taken to prevent contamination very low amount of bacteria may be obtained. Milk of better quality might be obtained with machine milking than by hand milking, because in that case the contamination from the hands of the milker and from the air may be prevented. In machine milking special care must be taken to sterilize the apparatus properly after each milking.

Milking Utensils

The milking utensils are good sources of contaminations of milk. There are two types of possibilities of contamintaion from this source. It may arise from the unsterilised milk utensils in which the type of bacteria may be lactic acid organisms or it may arise from the water with which the utensil was washed.

The type of contamination from the water may be of the spore forming bacteria. For lowering the number of bacteria in milk the milking utensils must be washed with hot water can be sterilized by steam. Most of the contamination in milk takes place after it is milked. Just after milking it contains much less bacteria and by the time it is bottled, the quantity of bacteria increases enormously.

Table. Contamination of Milk After Production.

Sl.No	Milk stage	No. of bacteria in 1 cu cm of milk
1	Milk after milking	19,000
2	After putting in the first pail	28,000
3	After pouring in the refrigerator	38,000
4	After pouring in the second pail	78,000
5	After pouring in bottles	1,62,000

Feeds

Since, bacteria are found on vegetation and in the soil, they are abundant in most of the cattle feeds. The feeding of excessive dust or mouldy feeds should be avoided if it is economically possible. Microflora of dry feeds consist of mostly aerobic spore forming bacteria. If silage is given to the cattle great care should be taken not to contaminate milk because of its liquid content and it contains very great quantity of streptococci and lactobacilli.

Air

The air of the barn is usually heavily contaminated during mixing of dry feeds and cleaning the floor. Generally the air contains very little quantity of micro-organisms.

Insects and rodents

Dairy plants must be kept free of insects and rodents. Few living things, other than micro-organisms, are threat to human health like flies, cockroaches, rats and mice. Their habits are unscrupulously filthy. They enter processing areas and equipments whenever possible. The bacterial count of the body of one fly may be many millions.

Water

Water supplies need special care of treatment to assure a quality suitable for cleaning and processing operations in a dairy or food processing plant. Water quality will vary with the source of supply. Water from surface supplies, such as a stream, pond or tank, is contaminated by dust, animals, plants, people, and other agents. Such water needs special treatment before it may be used in a dairy plant.

4

Milk Testing and Alcohol Test

RELEVANT INFORMATION

The alcohol test is used for rapid assessment of stability of milk for processing particularly for condensing and sterilization. The alcohol test is useful as an indication of the mineral balance of milk and not as an index of developed acidity. The test aids in detection abnormal milk such as colostrums, milk from animals in late lactation, milk from animals suffering from mastitis and milk in which mineral balance has been disturbed.

Precaution

Avoid direct heating of milk,

Material Required:

- Milk sample,
- 68% Ethyl alcohol by weight (Density 0.8675 g/mi. at 27 0 C).

Apparatus:

1. Test tube 15 x 1.9 cm preferably with graduation mark at 5 and 10 ml.
2. Measure of alcohol for 5 ml.
3. Test tube stand.

Procedure:

1. Take 5 ml. of milk in test tube.
2. Add equal quantity of 68% Ethyl alcohol.
3. Mix the contents of the test tube by inverting several times.
4. Examine the tube and note any coagulation.

If coagulation has occurred fine particles of curd will be visible on the inside surface, presence of flake or curd denotes positive alcohol test. Such samples are rejected.

ALCOHOL TEST FOR ASSESSING THE RAW MILK QUALITY

The alcohol test is used on fresh milk to indicate whether it will coagulate on thermal processing. This test is especially important for the manufacture of

UHT milk, evaporated milk and milk powders. This test is more sensitive than Clot-on Boiling (COB) test. It is based on tendency of milk protein to get unstable as a result of disturbance in the mineral balance of milk.

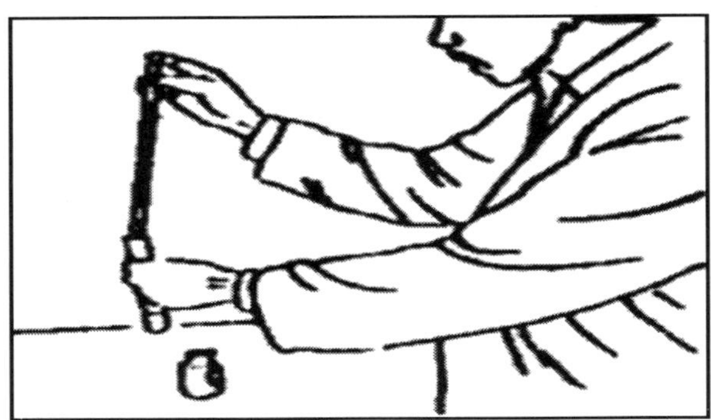

Fig. Alcohol Test for Assessing the Raw Milk Quality.

Milk with high developed acidity, or having calcium and magnesium compounds in greater than normal amounts, will coagulate when alcohol is added. Increased levels of albumen (colostrum milk) and salt concentrates (mastitis) may also results in a positive test. In dairy industries, normally three different concentrations of ethanol solution are used for the test, depending upon the further use of milk. These are 68% v/v, 65% v/v and 60% v/v. Milk which passes 68% ethanol test is considered as that of superior quality. For manufacturing UHT milk and milk powders, the raw milk should pass 68% ethanol test.

Dairies generally consider 60% ethanol test negative raw milk for manufacturing pasteurized polypack milk as it does not have to undergo rigorous heat treatment as in case of UHT milk. The milk which do not even pass 60% ethanol test is rejected at the processing units. Procedure: For routine testing, 5 ml milk is mixed with 5 ml of ethanol solution. If the tested milk is of good quality, there will be no coagulation, clotting or precipitation. Presence of flakes or clots indicates poor quality milk.

ALKALINE PHOSPHATASE TEST FOR PASTEURIZED MILK

Pasteurisation is an essential process in the production of milk which is safe and free from pathogens. Alkaline Phosphatase is an enzyme which is naturally present in milk, but is destroyed at a temperature just near to the pasteurization temperature. Alkaline Phosphatase test is used to indicate whether milk has been adequately pasteurised or whether it has been contaminated with raw milk after pasteurisation. This test is based on the principle that the alkaline phosphatase enzyme in raw milk liberates phenol from a disodium para-nitro phenyl phosphate and forms a yellow coloured complex at alkaline pH (Scharer, 1943). The intensity of yellow colour produced is proportional to the activity of the enzyme. The colour intensity is measured by direct comparison with standard colour discs in a Lovibond comparator. The test is not applicable to sour milk and milk preserved with chemical preservatives.

Apparatus required

- Water-Bath -maintained at 37±lpC, thermostatically controlled.
- Comparator - with special discs of standard colour glasses calibrated in μg p-nitrophenol per ml milk, and 2 × 25 mm cells.
- Test Tubes - of size 16 x 1.50 mm and rubber stoppers to fit.
- Pipettes - 1, 5, and 10 ml.
- Filter Paper - Whatman No. 2 or equivalent.
- Litmus Paper

Reagants

1. Sodium Carbonate-Bicarbonate Buffer - Dissolve 3.5 g of anhydrous sodium carbonate and 1.5 g of sodium bicarbonate in one litre of distilled water.
2. Buffer Substrate - Dissolve 1.5 g of disodium p-nitrophenyl phosphate in one litre of sodium carbonate-bicarbonate buffer. This solution is stable if stored in a refrigerator at 4°C or less for one month but a colour control test should be carried out on such stored solutions

Procedure

1. Pipette 5 ml of buffer substrate into a clean, dry test tube followed by 1 ml of the milk to be tested. Stopper the tube, mix by inversion and place in the water-bath
2. At the same time place in the water-bath a control tube containing 5 ml of the buffer substrate and 1 ml of boiled milk of the same kind as that under test that is pasteurized homogenized, low fat.
3. After 2 hours, remove the tubes from the bath, invert each and read the colour developed using the comparator and special disc, the tube containing the boiled milk control being placed on the left of the stand and the tube containing the sample under test on the right. Record readings which lie between two standard colour discs by adding a plus (+) or minus (-) sign to the figure of the nearest standard.

THE EFFICIENCY OF PASTEURISATION – THE PHOSPHATASE TEST

The Phosphatase test is a chemical measure for the efficiency of milk processing, revealing whether the products have been adequately heat-treated and the possible causes of under-pasteurisation or whether they have been contaminated with raw products. The Phosphatase test is valuable as a rapid test, designed to detect minor processing faults after an incubation period as short as 30 minutes, compared with incubation of up to 24 hours for alternative methods. It is used widely as a routine test in dairy plant control and has been adopted as the official test method in the UK and France.

QUALITY TEST BASED ON THE RESAZURIN REDUCTION METHOD

The Resazurin test is used as a rapid indication of the bacterial content of milk. Resazurin gives milk a characteristic blue colour and the test is based on the ability of bacteria in the milk to reduce the blue dye. The quality of the milk is judged by noting the degree of colour change – from blue through mauve and purple and pink and finally colourless – after a stated period of incubation, or the time required reducing the dye to a predetermined colour.

The Resazurin test is ideal for testing milk at the point of delivery and for regular periodic checks on quality. Although it is most applicable to raw milk, it has also been used to test pasteurised milk, raw and pasteurised cream, foremilk samples for detecting mastitis plus the cleanliness of shipping cans and other containers.

ESTIMATING SHELF-LIFE BY MEASURING DEVELOPED LACTIC ACID

This test for milk and milk products is based on a relationship between developed lactic acid and keeping quality. Milk proteins are precipitated by means of reagents and ferric chloride is added to the filtrate to produce a yellow complex. The intensity of this yellow colour is proportional to the concentration of lactic acid present.

Equipment Required: Each of these colorimetric tests can be carried out using a selection of inexpensive and simple to use equipment:

- The Lovibond® Comparator 2000 in conjunction with the
- Milk Viewing Stand, DB415,
- The appropriate reagents and
- Test discs containing coloured, graded glass filters and
- Test tubes, AF215TT.
- Also available are portable and laboratory lighting units for accurate colour matching irrespective of ambient lighting.

OTHER CHEMICAL ANALYSIS TESTS:

Alizarol (68%) Test

This test gives an indication of the heat stability of milk and is used for testing the stability of the milk proteins before processing. The test is used as a platform (acceptance/rejecting) test for raw milk for acceptability either on the farm, tanker or milk reception at the factory.

Results:

In summary, colour standards indicate:

1. Light purple (fresh milk)
2. Yellow (sour)
3. Brownish-yellow (slightly acid)
4. Violet (alkaline)

Lactognost Test

Qualitative analysis of phosphatase in milk, whey, cream and butter

Kit with dosed reagents for determination of alkaline phosphatase activity (limit test). The test enables the determination of H.T.S.T. heating of milk and similar products or the determination of raw milk in pasteurized milk.

MILK TESTING - PLATFORM TESTS

RELEVANT INFORMATION

Platform tests include the tests for judging the quality of the raw milk.

These are:

- Organoleptic evaluation (OE),
- Clot on boiling test (COP),
- Alcohol test (AT),
- Sediment test (ST)
- Resazurin test (RT).

The milk is collected from various sources and transported to milk scheme for processing, marketing and distribution. Large quantity of milk is supplied to the plant through different agencies, so that is subjected to check for its suitability. Hence it is essential to examine the milk by using different platform tests.

Precaution:

1. Raw milk should be tested.
2. Curdled/ spoiled milk should be separated.
3. Temperature of milk while receiving at milk plant should be 50 C.

Material Required: Milk cans or containers. Apparatus:

1. Thermometer.
2. Dipper
3. Plunger.

Procedure: Organoleptic Evaluation: Examine the given sample of milk and record your observations on following aspects. Odour/ Smell:

1. It can be judged within few seconds.
2. Remove the lid of can hold it inverted and raised to the nose and inhale the smell.
3. Record the odour/ smell as normal or abnormal.
4. Milk should be free from any off flavour like feed, fishy, barny etc.

General Appearance

Note whether the milk is clear or any visible dirt or foreign matter has gained entrance in milk. If so describe its nature for detecting the amount of dirt, in the milk, conduct the sediment test if necessary.

Colour

Observe the colour of milk as white, light yellow; record whether colour of milk is normal or abnormal. (Abnormal colours are reddish, bloody, bluish etc.)

Consistency

Record the consistency of milk as normal, watery, thick, ropy, and slimy.

Temperature

Note down the temperature of milk at the time of receiving. It should be below 500 F.

MILK TESTING - CLOT ON BOILING TEST (COB)

Objectives: To determine the stability of milk for heat processing.

Relevant Information: If milk is kept as such at room temperature, there will be increased in the acidity which is called as developed acidity. If acidity is increased to more than 0.2 percent, there is coagulation due to heat treatment, which is the result of dissociation of calcium caseinate salt. Hence it is essential to know the heat stability of incoming raw milk for further processing.

Precaution: Avoid direct heating of milk. Use hot water bath.

Material Required:

1. Milk and milk container.

Apparatus:

1. Test tube 20 ml. capacity.
2. Hot water bath
3. Test tube stand
4. Heating arrangement (Electrical connection).

Procedures:

1. Take 5 ml of milk in the test tube.
2. Put this on boiling water bath for 5 minutes.
3. Remove the tube from water bath without shaking.
4. Note any acid smell or precipitated particles on the sides of the test tube.

Sample showing precipitated particles are recorded as positive C.O.B. test. Such milk is rejected on the platform.

MILK TESTING - TWO MINUTES RESAZURIN TEST

Relevant Information

The majority of the organisms in milk are capable of reducing and decolorising the resazurin dye. When bacteria grow in the milk they utilize oxygen, the rate of remove or reduction is proportional to the keeping quality. This test is also based on the same principles as M.B.R. (Methylene Blue Reduction) test, but dye is Resazurin which is much more sensitive that the Methylene blue. For this reason this test provides a rapid measure of the keeping quality of milk.

During incubation, the dye undergoes reduction very largely through the metabolic activity of the organisms present. The greater the number of organisms present in milk, the more quickly the dye is reduced. The reduction takes place in two distinct stages. Resazurin is blue at the reaction of milk. In the first stage dye is changed to pink colour and in second stage pink colour is changed to colourness. The cells present in the milk may also influence the reduction of Resazurin and for that reason; the test may also measure physiologically or pathologically abnormal milk.

Precaution: The test should be carried out aseptically.

Material Required:

1. Milk sample
2. Resazurin colour/solution 0.05%.

Apparatus:

1. All purpose lovibond comparator.
2. One Resazurin colour disc from blue to white.
3. Water path, thermostically controlled to maintain temperature of 370 C.
4. Test tube 10 ml.
5. Pipette 10 ml. and 1 ml.

Procedure:

- Milk the sample thoroughly by inverting from one to another container.
- Pour 10 ml. of milk sample in to previously sterilized test tube.
- Add quickly 1 ml. of Resazurin solution in the test tube.
- Mix the milk and dye thoroughly by inverting 2-3 times.
- Place the tubes in the water bath at the temperature of 37.5 0 C only for two minutes.
- The tubes are then removed from the water bath.
- Compare the colour to test tube with standard disc until the colours are matched under comparator.
- Record the number of disc/colour of disc. If colour falls between two disc numbers record half value.

MILK TESTING - SEDIMENT TEST

RELEVANT INFORMATION

Sediment test of raw milk will reveal the extent to which visible insoluble matter has gained entrance in the milk. It is a rapid test indication quantitative measure of carelessness in handling the milk and lack of sanitation. But in milk that appears as visible or insoluble sediment is always associated with relative amount of microbes. The test is carried out by allowing a measured quantity of milk to pass through a fixed area of a filter disc and comparing the sediment with the prescribed standards.

Precautions:

1. Row milk should be tested.
2. Curdled milk should be separated.
3. Temperature of milk while receiving at milk plant should be 50 C.
   Material Required: Milk

Apparatus:

- Sediment tester:This apparatus consists of a bottle open at both ends. To the neck of which is fastened a clip and wire guage. This inverted neck is kept downwards in stand small cotton discs are used to retain the dirt.
- Lintine disc 32 mm in diameter.
- Sampling dipper of 500 ml. capacity.
- Sediment disc ratings: 0.0, 0.2, 0.5, 1.0, 2.0 mg sediment per 500 ml. of milk.
- Sieves

Procedure:

1. Take a milk sample from well stirred can of milk with the help of sampling dipper.
2. Filter the milk through properly adjusted firm link disc held in the sediment tester, so that a filtration area of 28 mm in diameter is exposed.
3. Remove the cotton disc from sediment tester after filternation.

SMALL DAIRY INDUSTRY: MILK SEDIMENT TESTING

According to the United States Department of Agriculture (USDA) National Agricultural Statistics Service (NASS), milk as a commodity was worth $39 billion in 2011. As of 2009, NASS reported that the trend in dairy was toward large scale operations, defined as a milk cow ranch with greater than 500 head of cattle. However, in 2009, more than 65,000 small dairies in the US were still operating. The small dairies are less likely to use fully automated clean-in-place machinery to prevent contamination. In 2010, the USDA rescinded its requirement that large-scale dairies test their milk for sediment contamination, but smaller dairies collecting milk in cans must continue to follow established guidelines. These small-scale dairies must prove to authorities that their product is clean by allowing milk inspectors to perform official milk sediment testing once per month. Still, the owner of a small dairy may want to perform their own sediment testing more frequently to address contamination issues before the inspector arrives.

PATHOGENIC MICROBES IN HUMAN ILLNESS

Before the turn of the century, the role of pathogenic microbes in human illness was discovered by Dr. Robert Koch. Dr. Koch and his contemporaries began developing testing methods for the detection of pathogens in water, food, and other products that came in contact with the general public.

With regard to milk, the presence of sediment was highly correlated to bacterial contamination. By 1914, there were twelve widely used methods for detecting contamination in milk alone, falling into two groups: filtration and centrifugation based methods. Although pasteurization would kill off the bacteria, the byproducts of the contamination remain, causing foul flavors and odors. If milk contains too much sediment, it is rejected by milk inspectors, and steps must be taken to prevent sediment contamination before the supplier's milk will be deemed suitable for public sale.

In 1952, the Association of Official Agricultural Chemists (AOAC) published a method adopted by the USDA known as the Milk Sediment Test. It uses a special vacuum filtration apparatus and specific filter papers to collect the sediment. The captured sediment is then graded on a qualitative scale by comparing the sediment disk against the USDA photographic standards.

The Milk Sediment Test requires several parts, including the sediment testing apparatus, sediment filter papers, a vacuum filter flask, a vacuum pump, stopper, and tubing. Some sediment testers have an integrated aspirator pump (a type of vacuum pump). One of the most common variations of the method requires that a one-gallon sample be filtered through a 1.25 inch filter. The milk sediment testing apparatus has been designed to prohibit any milk from bypassing the filter paper. The vacuum pump pulls the milk through the filter at a much faster rate than gravity alone can accomplish, so testing each bulk tank at least three times, as proscribed by the AOAC method, doesn't take most of a day. Milk sediment filter papers are available in white and black, and it is recommended to use both. The white paper works well for black or colored particulates. The black paper is best for white or clear particulates.

Once the AOAC method becomes part of normal operations, it can be used to identify sources of contamination at every step in the process, from poor cow hygiene to inadequately washed containers to improper cleaning of milking stalls. Milk sediment disk analysis is easily performed by stereo or bright field compound microscopy, because particulates can often be identified by their appearance.

OPERATING PROCEDURE OF MILK PROCUREMENT

To meet out the milk requirement for manufacturing different milk products, VRS Foods Limited procures good quality milk from 3 sources:

THROUGH FRANCHIS

In this procedure we provide our Infrastructure for milk chilling, staff for grading & testing of milk and tankers for timely lifting of milk from MCC, to a person who procure milk on our behalf. Milk comes directly to our MCC dock, where after proper grading and testing milk is accepted.

Through Supplier

In this system, persons who have their own chilling facilities, procure milk through small vendors and after proper chilling of milk, send the milk to our dairies through their own tankers. At dairy, after proper testing on different parameters, if milk is found as per prescribed norms is accepted.

Through VLC System (Direct from Producers at Village level)

Salient features of this system are as follows:

- Procurement and collection of milk is undertaken with most modern management systems and IT intervention. It involves use of Automatic milk collection Unit (AMCU/DPMCU). The person is selected as VSP after proper screening.
- The milk is collected in both shifts i.e. morning and evening throughout the year, through the village level milk collection centers. (VLCs).
- Each VLC is staffed and equipped with the following: One educated person called village service provider (VSP), having a minimum qualification of matriculation, from the same village, who is adequately trained in clean milk production, collection and testing of milk,

dispatching and transportation of milk, operating AMCU/DPMCU, pricing of milk, record keeping and reporting etc.

- Each VSP is required to have adequate space for collection and testing of milk, which can accommodate testing equipments, milk collection and transporting milk cans and keeping records etc.
- Each producer is trained in hygienic milking procedure and carrying milk in a steal bucket, with a cover on it to VLC.
- A team of qualified and trained supervisors is engaged to supervise & monitor the activities at VLC level. They also train, guide and facilitate the VSP to accomplish his task of procuring good quality milk from the producers.
- Milk from VLCs is brought to MCC through milk vans in 40 litre aluminum cans, as per the schedule time table so that quality of milk remains good and it is immediately chilled below 40C, to keep the quality of milk intact.
- At dairy dock of MCC, qualified staff is deputed for proper grading and testing of milk and as per norms, only good quality of milk is accepted.
- VRS Foods Ltd. Is already catering the demand of urban masses of milk & milk products through a net work of distributors and through door-to-door delivery system, in NCR, Delhi and western part of UP.
- From Dairies as per the demand of market, milk & milk products are supplied through insulated vans so that quality of milk & milk products could be maintained as per norms and Govt. guidelines.
- The left out urban areas would be covered through direct selling of milk & milk products through distributors and vendors' network.

PROCESSING & LOGISTICS

To maintain the quality of milk procured from the rural areas approximately 100 milk chilling centers would be hired in the area of operation (In the 5th year of operation) so that within 2.30 hrs milk would be chilled afterprocurement.VRS Foods Ltd. Is having its processing units at Sahibabad, Gulaothi, Sandila and malanpur for providing milk and milk products to urban masses.

These dairies will cater liquid milk and indigenous milk products demand of neighboring districts. These dairies are meeting the demand of butter, ghee, cheese and export quality milk products. From the dairies, milk and milk products would be transported through insulated vans to neighboring districts so that quality of milk & milk products could be maintained as per prescribed norms and guidelines.

MILK TESTS FOR BRUCELLOSIS

BRBME - BRUCELLOSIS BULK MILK ELISA (ENZYME LINKED IMMUNOSORBENT ASSAY)

The test procedure is known as the Enzyme Linked ImmunoSorbent Assay, or ELISA for short. This test type may be used to detect a number of diseases in humans and animals using blood serum or milk. When used to test milk for brucellosis we know it as the Brucellosis Bulk Milk ELISA or BrBME. It has been used in Great Britain and in the Republic of Ireland for some time to detect brucellosis and is capable of detecting one infected animal in a herd of 100 - 200 milking cows.

As with any other biological test, there may be false positive and false negative results. In other words not every herd reacting positively to the test is infected, and not every infected herd will give a positive reaction to the test. For this reason the test is not considered to be a definitive indicator of disease. Rather, a positive or inconclusive result is an indicator that individual blood testing of the herd should be carried out to determine if an infected animal is present.

BRUCELLOSIS

Brucellosis, Bang's disease, Crimean fever, Gibraltar fever, Malta fever, Maltese fever, Mediterranean fever, rock fever, orundulant fever, is a highly contagious zoönosis caused by ingestion of unpasteurized milk or undercooked meat from infected animals or close contact with their secretions.

Brucella species are small, Gram-negative, nonmotile, nonspore-forming, rod-shaped (coccobacilli) bacteria. They function asfacultative intracellular parasites, causing chronic disease, which usually persists for life. Four species infect humans: B. melitensis,B. abortus, B. suis, and B. canis. B. melitensis is the most virulent and invasive species; it usually infects goats and occasionally sheep. B. abortus is less virulent and is primarily a disease of cattle. B. suis is of intermediate virulence and chiefly infects pigs. B. canis affects dogs. Symptoms include profuse sweating and joint and muscle pain. Brucellosis has been recognized in animals and humans since the 20th century.

Signs and Symptoms

The symptoms are like those associated with many other febrile diseases, but with emphasis on muscular pain and sweating. The duration of the disease can vary from a few weeks to many months or even years. In the first stage of the disease, septicaemia occurs and leads to the classic triad of undulant fevers, sweating (often with characteristic smell, likened to wet hay), and migratory arthralgia and myalgia (joint and muscle pain). Blood tests characteristically reveal leukopenia andanemia, show some elevation of AST and ALT, and demonstrate positive Bengal Rose and Huddleston reactions. This complex is, at least in Portugal, Israel, and Jordan, known as Malta fever. During episodes of Malta fever, melitococcemia (presence of brucellae in blood) can usually be demonstrated by means of blood culture in tryptose medium or Albini medium. If untreated, the disease can give origin to focalizations or become chronic. The focalizations of brucellosis occur usually in bones and joints and spondylodiscitis of the lumbar spine accompanied by sacroiliitis is very characteristic of this disease. Orchitis is also common in men.

Diagnosis of brucellosis relies on:

- *Demonstration of the agent:* blood cultures in tryptose broth, bone marrow cultures. The growth of brucellae is extremely slow (they can take up to two months to grow) and the culture poses a risk to laboratory personnel due to high infectivity of brucellae.
- Demonstration of antibodies against the agent either with the classic Huddleson, Wright, and/or Bengal Rose reactions, either with ELISA or the 2-mercaptoethanol assay for IgM antibodies associated with chronic disease
- Histologic evidence of granulomatous hepatitis on hepatic biopsy
- Radiologic alterations in infected vertebrae: the Pedro Pons sign (preferential erosion of the anterosuperior corner of lumbar vertebrae) and marked osteophytosis are suspicious of brucellic spondylitis.

The disease's sequelae are highly variable and may include granulomatous hepatitis, arthritis, spondylitis, anaemia, leukopenia, thrombocytopenia, meningitis, uveitis, optic neuritis, endocarditis, and various neurological disorders collectively known as neurobrucellosis.

Cause

Brucellosis in humans is usually associated with the consumption of unpasteurized milk and soft cheeses made from the milk of infected animals, primarily goats, infected with Brucella melitensis and with occupational exposure of laboratory workers, veterinarians, and slaughterhouse workers. Some vaccines used in livestock, most notably B. abortus strain 19, also cause disease in humans if accidentally injected. Brucellosis induces inconstant fevers, miscarriage, sweating, weakness, anaemia, headaches, depression, and muscular and bodily pain. The other strains are Brucella suis and Brucella canis, which cause infection in pigs and dogs respectively.

Diagnosis

Definite diagnosis of brucellosis requires the isolation of the organism from the blood, body fluids or tissues, but serological methods may be the only tests available in many settings. Positive blood culture yield ranges between 40% and 70% and is less commonly positive forB. abortus than Brucella melitensis or B. suis. Identification of specific antibodies against bacterial lipopolysaccharide and other antigenscan be detected by the standard agglutination test (SAT), rose Bengal, 2-mercaptoethanol (2-ME), antihuman globulin (Coombs') and indirect enzymelinked immunosorbent assay (ELISA). SAT is the most commonly used Serology in endemic areas. An agglutination titre of e"1: 160 is considered significant in non-endemic areas and e"1: 320 in endemic areas. Due to the similarity of the O. polysaccharide of Brucella to that of various other Gram-negative bacteria (e.g. Francisella tularensis, Escherichia coli, Salmonella urbana, Yersinia enterocolitica, Vibrio cholerae, and Stenotrophomonas maltophilia) the appearance of cross-reactions of class M immunoglobulins may occur.

The inability to diagnose B. canis by SAT due to lack of cross-reaction is another drawback. False-negative SAT may be caused by the presence of blocking antibodies (the prozone phenomenon) in the á2-globulin (IgA) and in the á-globulin (IgG) fractions. Dipstick assays are new and promising, based on the binding ofBrucella IgM antibodies, and found to be simple, accurate and rapid. ELISA typically uses cytoplasmic proteins as antigens. It measures IgM, IgG and IgA with better sensitivity and specificity than the SAT in most recent comparative studies. The commercial Brucellacapt test, a single-step immunocapture assay for the detection of total anti Brucella antibodies, is an increasingly used adjunctive test when resources permit. PCR is fast and should be specific. Many varieties of PCR have been developed (e.g. nested PCR, realtime PCR and PCR-ELISA) and found to have superior specificity and sensitivity in detecting both primary infection and relapse after treatment. Unfortunately, these have yet to be standardized for routine use, and some centres have reported persistent PCR positivity after clinically successful treatment, fuelling the controversy about the existence of prolonged chronic brucellosis. Other laboratory findings include normal peripheral white cell count, and occasional leucopenia with relative lymphocytosis. The serum biochemical profiles are commonly normal.

Prevention

United States

Dairy herds in the USA to be Certified Brucellosis-Free are tested at least once a year with the Brucella milk ring test (BRT). Cows confirmed to be infected are often killed. In the United States, veterinarians are required to vaccinate all young stock, thereby further reducing the chance of zoonotic transmission. This vaccination is usually referred to as a "calfhood" vaccination. Most cattle receive a tattoo in their ear serving as proof of their vaccination status. This tattoo also includes the last digit of the year they were born. The first state–federal cooperative efforts towards eradication of brucellosis caused by Brucella abortus in the U.S. began in 1934. Wild bison and elk in the Greater Yellowstone Area (GYA) are the last remaining reservoir of B. abortus in the US. The recent transmission of brucellosis from elk to cattle in Idaho and Wyoming illustrates how the GYA, as the last remaining reservoir in the United States, may adversely affect the livestock industry. Eliminating brucellosis from this area is a challenge, as many viewpoints exist on how to manage diseased wildlife.

Canada

The Canadian government declared its cattle population to be brucellosis-free on 19 September 1985. The brucellosis ring testing of milk and cream, as well as the testing of cattle to be slaughtered, ended on 1 April 1999. Monitoring continues through testing at auction markets, through standard disease-reporting procedures, and through the testing of cattle being qualified for export to countries other than the United States.

Europe

Malta

Until the early 20th century, the disease was endemic in Malta to the point of it being referred to as "Maltese fever". The link between the illness and unpasteurised milk was established by Temi Zammit. Today, due to a strict regimen of certification of milk animals and widespread use of pasteurization, the illness has been eradicated from Malta.

Republic of Ireland

Ireland was declared free of brucellosis on 1 July 2009. The disease had troubled the country's farmers and veterinarians for several decades. The Irish government submitted an application to the European Commission, which verified that Ireland had been liberated. Brendan Smith, Ireland's then Minister for Agriculture, Food and the Marine, said the elimination of brucellosis was "a landmark in the history of disease eradication in Ireland". Ireland's Department of Agriculture, Food and the Marine intends to reduce its brucellosis eradication programme now that eradication has been confirmed.

Australia

Australia is free of cattle brucellosis, although it occurred in the past. Brucellosis of sheep or goats has never been reported. Brucellosis of pigs does occur. Feral pigs are the typical source of human infections.

New Zealand

Brucellosis in New Zealand is limited to sheep (B. ovis). The country is free of all other species of Brucella.

Treatment

Antibiotics such as tetracyclines, rifampicin, and the aminoglycosides streptomycin and gentamicin are effective againstBrucella bacteria. However, the use of more than one antibiotic is needed for several weeks, because the bacteria incubate within cells. Surveillance using serological tests, as well as tests on milk like the milk ring test, can be used for screening and play an important role in campaigns to eliminate the disease. As well individual animal testing both for trade and for disease control purposes is practiced. In endemic areas, vaccination is often used to reduce the incidence of infection. Several vaccines are available that use modified live viruses. The World Organisation for Animal Health Manual of Diagnostic Test and Vaccines for Terrestrial Animals provides detailed guidance on the production of vaccines.As the disease is closer to being eliminated, a test and stamping out program is required to completely eliminate it.

The gold standard treatment for adults is daily intramuscular injections of streptomycin 1 g for 14 days and oral doxycycline 100 mg twice daily for 45 days (concurrently). Gentamicin 5 mg/kg by intramuscular injection once daily for seven days is an acceptable substitute when streptomycin is not available or contraindicated. Another widely used regimen is doxycycline plus rifampin twice daily for at least six weeks. This regimen has the advantage of oral administration. A triple therapy of doxycycline, with rifampin and co-trimoxazole, has been used successfully to treat neurobrucellosis.

Doxycycline is able to cross the blood–brain barrier, but requires the addition of two other drugs to prevent relapse. ciprofloxacin and co-trimoxazole therapy is associated with an unacceptably high rate of relapse. In brucellic endocarditis, surgery is required for an optimal outcome. Even with optimal antibrucellic therapy, relapses still occur in 5 to 10% of patients with Malta fever.

The main way of preventing brucellosis is by using fastidious hygiene in producing raw milk products, or by pasteurizing all milk that is to be ingested by human beings, either in its unaltered form or as a derivate, such as cheese. Co-trimoxazole and rifampin are both safe drugs to use in treatment of pregnant women who have brucellosis.

Prognosis

The mortality of the disease in 1909, as recorded in the British Army and Navy stationed in Malta, was 2%. The most frequent cause of death was endocarditis. Recent advances in antibiotics and surgery have been successful in preventing death due to endocarditis. Prevention of human brucellosis can be achieved by eradication of the disease in animals by vaccination and other veterinary control methods such as testing herds/flocks and slaughtering animals when infection is present. This requires prolonged commitment and financial support at the highest levels of government, with strong cooperation between both animal and public health authorities. Currently, no effective vaccine is available for humans. Boiling milk before consumption, or before using it to produce other dairy products, is protective against transmission via ingestion. Changing traditional food habits of eating raw meat, liver or bone marrow is necessary but difficult to implement. Patients who have had brucellosis should probably be excluded indefinitely from donating blood or organs. Exposure of diagnostic laboratory personnel to Brucella organisms remains a problem in both endemic settings and when brucellosis is unknowingly imported by a patient. After appropriate risk assessment, staff with significant exposure should be offered post-exposure prophylaxis (PEP) and followed up serologically for 6 months. Recent published experience confirms that prolonged and frequent serological follow-up consumes significant resources without yielding much information, and is burdensome for the affected staff, who often fail to comply. The side effects of the usual recommended regimen of rifampicin and doxycycline for 3 weeks also reduce treatment adherence. As there is no evidence that PEP with two drugs is superior to monotherapy, British guidelines now recommend PEP with doxycycline alone for 3 weeks and a less onerous follow-up protocol.

History

Under the name "Malta fever", the disease now called brucellosis first came to the attention of British medical officers in the 1850s in Malta during the Crimean War. Jeffery Allen Marston (1831-1911) described his own case of the disease in 1861. The causal relationship between organism and disease was first established in 1887 by David Bruce.

In 1897, Danish veterinarian Bernhard Bang isolated Brucella abortus as the agent; and the additional name "Bang's disease" was assigned. Maltese doctor and archaeologist Sir Themistocles Zammit earned a knighthood for identifying unpasteurized milk as the major source of the pathogen in 1905, and it has since become known as Malta fever. In cattle, this disease, usually caused by B. abortus, is also known as "contagious abortion" and "infectious abortion".

The popular name "undulant fever" originates from the characteristic undulance (or "wave-like" nature) of the fever, which rises and falls over weeks in untreated patients. In the 20th century, this name, along with brucellosis (after Brucella, named for Bruce), gradually replaced the 19th century names Mediterranean fever and Malta fever. In 1989, neurologists in Saudi Arabia discovered "neurobrucellosis", a neurological involvement in brucellosis.

The following obsolete names have previously been applied to brucellosis:

- Backdoor trauma
- Brucelliasis

- Bruce's septicemia
- Chumble fever
- Contagious abortion
- Continued fever
- Crimean fever
- Cyprus fever
- Febris melitensis
- Febris undulans
- Fist of mercy
- Five dollar disease
- Goat fever
- Jones disease
- Maltese fever
- Melitensis septicemia
- Melitococcosis
- Milk sickness
- Mountain fever
- Neapolitan fever
- Satan's fever
- Scottish delight
- Slow fever

Biological Warfare

Brucella species were weaponized by several advanced countries by the mid-20th century. In 1954, B. suis became the first agent weaponized by the United States at its Pine Bluff Arsenal near Pine Bluff, Arkansas. Brucella species survive well in aerosols and resist drying. Brucella and all other remaining biological weapons in the U.S. arsenal were destroyed in 1971–72 when the American offensive biological warfare program was discontinued by order of President Richard Nixon.

The experimental American bacteriological warfare program focused on three agents of the Brucella group:

- Porcine brucellosis (agent US)
- Bovine brucellosis (agent AB)
- Caprine brucellosis (agent AM)

Agent US was in advanced development by the end of World War II. When the U.S. Army Air Forces (USAAF) wanted a biological warfare capability, the Chemical Corps offered Agent US in the M114 bomblet, based on the four-pound bursting bomblet developed for spreading anthrax during World War II. Though the capability was developed, operational testing indicated the weapon was less than desirable, and the USAAF designed it as an interim capability until it could replaced by a more effective biological weapon. The main drawbacks of the M114 with Agent US was that it was an incapacitating agent, whereas the administration of the USAAF wanted deadly weapons. Also, the stability under storage was too low to allow for storing at forward air bases, and the logistical requirements to neutralize a target were far higher than originally planned. This would have required an unreasonable amount of logistical support.

Agents US and AB had a median infective dose of 500 organisms/person, and for Agent AM it was 300 organisms/person. The time-of-incubation was believed to be about two weeks, with a duration of infection of several months. The lethality estimate was based on epidemiological information at 1 to 2%. Agent AM was believed to be a more virulent disease, and a fatality rate of 3% was expected.

Other Animals

Species infecting domestic livestock are B. melitensis (goats and sheep), B. suis (pigs), B. abortus (cattle and bison), B. ovis (sheep), and B. canis (dogs). B. abortus also infects bison and elk in North America and B. suis is endemic in caribou. Brucella species have also been isolated from several marine mammal species (pinnipeds and cetaceans).

Cattle

Brucella abortus is the principal cause of brucellosis in cattle. The bacteria are shed from an infected animal at or around the time of calving or abortion. Once exposed, the likelihood of an animal becoming infected is variable, depending on age, pregnancy status, and other intrinsic factors of the animal, as well as the number of bacteria to which the animal was exposed. The most common clinical signs of cattle infected with B. abortus are high incidences of abortions, arthritic joints, and retained placenta.

The two main causes for spontaneous abortion in animals are erythritol, which can promote infections in the fetus and placenta, and the lack of anti-Brucella activity in the amniotic fluid. Males can also harbor the bacteria in their reproductive tracts, namely seminal vesicles, ampullae, testicles, and epididymes.

Dogs

The causative agent of brucellosis in dogs, B. canis, is transmitted to other dogs through breeding and contact with aborted fetuses. Brucellosis can occur in humans who come in contact with infected aborted tissue or semen. The bacteria in dogs normally infect the genitals and lymphatic system, but can also spread to the eyes, kidneys, andintervertebral discs. Brucellosis in the intervertebral disc is one possible cause of discospondylitis. Symptoms of brucellosis in dogs include abortion in female dogs and scrotalinflammation and orchitis in males. Fever is uncommon. Infection of the eye can cause uveitis, and infection of the intervertebral disc can cause pain or weakness. Blood testing of the dogs prior to breeding can prevent the spread of this disease. It is treated with antibiotics, as with humans, but it is difficult to cure.

Cetaceans

Brucellosis is caused by the bacterium Brucella ceti. First discovered in the aborted fetus of a bottlenose dolphin, the structure of B. ceti is similar to Brucella in land animals. B. ceti is commonly detected in two suborders of cetaceans, Mysticetia and Odontoceti. Mysticetia includes four families of baleen whales, filter-feeders and Odontoceti includes two families of toothed cetaceans ranging from dolphins to sperm whales. It is believed that B. ceti is transferred from animal to animal through sexual intercourse, maternal feeding, aborted fetuses, placental issues, from mother to fetus or through fish reservoirs. Brucellosis is a reproductive disease and therefore has an extreme negative impact on the population dynamics of a species. This becomes even a greater issue when the already low population numbers of cetaceans are taken into consideration.

B. ceti has been identified in four out of the fourteen cetacean families, but the antibodies have been dedicated in seven of the families. This indicates that B. ceti is common amongst cetacean families and populations. Only a small percentage of exposed individuals become ill or die; yet this disease affects half of cetacean families. However, particular species, it appears, are more likely to become infected by B. ceti.

The harbor porpoise, striped dolphin, white-sided dolphin, bottlenose dolphin and the common dolphin are the species with the highest frequency of infection amongst ondontocetes. In the mysticetes family, the northern minke whale is by far the most infected species. Evidence indicates that dolphins and porpoises are more likely to be infected than cetaceans such as whales. With regard to sex and age biases, the infections do not seem to be influenced by the age or sex of an individual. Although fatal to cetaceans, B. ceti has a low infection rate for humans.

VALUE OF BRBME TESTS

Over four years from March 2000-2004, 254 non-negative herds were identified and subjected to blood testing. Of these, 106 had already been identified by other means as being at special risk and blood tests had been arranged. The remaining 148 were on a routine blood-testing regime when they showed non-negative to the BrBME. Of these, 41 were found to be infected with brucellosis. Thus the BrBME led to the detection of infected herds which might otherwise have been left to spread disease to their neighbours and further afield for up to two years before the presence of the disease would have been detected through routine testing. The test has therefore proved its worth beyond doubt.

BrBME Benefits

The main advantage of the BrBME is that the brucellosis status of all milking cows in a herd can be assessed by taking one milk sample from the bulk tank. This means cattle do not need to be individually restrained and tested. Indeed Department of Agriculture and Rural Development (DARD) staff do not even need to visit the farm as the milk samples are supplied by the dairy which collects milk from the farm. This makes sampling very cost effective. Because they are undergoing monthly BrBME, dairy herds in clear areas can be accorded a lower priority with regard to blood testing. This allows DARD staff to concentrate their efforts on the blood sampling of other herds. The test is recognised by the European Union as an official test for the purpose of the brucellosis control programme.

BrBME Drawbacks

The test does not, of course, identify suspect individuals within the herd. Indeed, it is not a definitive indicator of infection within the herd. However, it does highlight herds in which there may be a problem and allows us to conduct further investigations in these herds. A few herds throw up repeated positive results even though there is no infection in these herds.

Sample Collection for BrBME Tests

The vast majority of dairy farms have their milk collected by one of the large dairies which supply the Department of Agriculture and Rural Development (DARD) with monthly milk samples for Brucellosis testing. The majority of these draw off a sample of milk while the milk tanker pumps milk out of the bulk tank. These samples, contained in small snap-top containers, are sent by the Dairies to AFBI laboratories at Stormont.

Dairies based in the Republic of Ireland decant milk samples from their Northern Ireland suppliers into 'deep well blocks' which take up much less space than the original containers. These deep well blocks are then sent to the AFBI labs. A few dairy farms do not have their milk collected by the larger dairies. In such cases, DARD staff visit these farms monthly to collect samples which are then processed in the usual way.

Interpretation of BrBME Test Results

Results are provided by the lab to the Veterinary Service, and are categorised into negative, inconclusive, low positive, medium positive, and high positive. Currently, we take identical action in all cases where the result is 'non-negative'. With increasing experience in the use of this test, differentiation may be made. In the case of larger herds, the inherent dilution factor means that negative results must be regarded with caution.

Follow-up for non-Negative BrBME Results

A herd restriction is placed on all the herds giving positive or inconclusive readings to the Brucellosis Bulk Milk ELISA (Enzyme Linked ImmunoSorbent Assay)(BrBME). Veterinary Service staff will take a repeat sample from the bulk tank as a check on the original result. A herd blood test will be arranged and carried out as soon as possible. The results of this test are interpreted in the normal way and follow-up action is as for any other blood test.

Repeat BrBME Positive Results

Some herds yield positive results to bulk milk tests month after month despite the fact that blood tests reveal no sign of disease in the herd. The assumption is that this is caused by interference from antibodies to other diseases. While this situation continues, the test is, of course, of no value in detecting the introduction of Brucellosis to such herds. In such cases we may carry out individual milk sampling of cows in these herds, to identify the offending animals.

TARGET PROGESTERONE MILK TEST FOR BOVINE

This treatment applies to the following species:

- Beef Cattle
- Dairy Cattle

Manufacturer: BioMetallics

Rapid Test for Progesterone in Bovine Milk

Components: 20 monoclonal antibody treated test cups.

20 sample applicators.

1 bottle enzyme conjugate.

1 bottle wash solution.

1 bottle substrate buffer A.

1 bottle substrate buffer B.

1 substrate mixing bottle.

TARGET PROGESTERONE MILK TEST FOR BOVINE INDICATIONS

Recommended to confirm estrus, detect pregnancy, improve controlled breeding, define cycling, diagnose infertility disorders, and to monitor fertility treatment. Test Procedure: Milk Sample Collection: Milk samples should be obtained preferably from the foremilk or milk jug and put into a clean container. Last milk can be used, but is less preferred. Initial strippings should not be used. Label samples with the cow number sampling date, and cycle day. If a milk sample is not tested within 30 minutes it should be stored in the refrigerator. If the milk sample is to be tested more than 6 hours later, immediately after collection, add one drop of milk preservative (10% potassium dichromate) or 1/4 milk preservative tablet to each 10 mLs of milk. Store milk in the refrigerator - do not freeze. It is not recommended to test samples more than 4-5 days after collection.

Test Kit Preparation: Allow samples and all kit reagents to come to room temperature before starting test (2-24 hrs). Use a timer while performing the test. Label each test cup to be used with the cow number and line up samples with respective cups.

When to Use TARGET:

- *Confirm estrus (heat):* Take a milk sample on the day of suspected estrus. A bright blue test result indicates the cow is in estrus and ovulation is about to occur. The optimal time to inseminate is approximately 18 hours after obtaining a bright blue result.
- *Detect pregnancy or open cows:* Take a milk sample 19 to 21 days after insemination. A bright blue test result indicates the cow is open and is in estrus. A white test result strongly indicates the cow is pregnant. An additional sample taken at 22 or 23 days increases the accuracy of a pregnancy determination - up to 95% accuracy if a white result is obtained on both tests.

Pregnancy determination should be confirmed by a licensed veterinarian since, high progesterone could indicate mid ovarian cycle because of inaccurately timed insemination, or development of a persistent corpus luteum. Furthermore, cows diagnosed with high progesterone on day 19-23 may return to estrus if fetal reabsorption occurs. A milk sample taken at the time of palpation can be used in conjunction with palpation results.

- *Improve prostaglandin controlled breeding:* Take a milk sample prior to injection with prostaglandin. If the corpus luteum (CL) is functional (white result), continue with treatment. Prostaglandin injections after 50 days post-calving has been shown to reduce days open, reduce services per conception, and increase conception rate.
- *Define cycling:* Take milk samples at 3 to 5 day intervals for a period of 3 weeks or more. A combination of low and high progesterone results (any shade of blue and white) indicates that ovarian activity has resumed (usually evident about 1 month post-calving).
- *Diagnose infertility disorders:* After veterinarian confirmation of the presence of an ovarian cyst (by rectal palpation), target is useful in determining the type of cyst. Take one milk sample at the day of palpation and 2 more samples at 7 day intervals for questionable cases.

A bright blue test result (low progesterone) strongly indicates a follicular cyst or nonfunctional ovary. A white test result (high progesterone) strongly indicates the presence of a luteal cyst or cystic CL.

- Monitor fertility treatment: Take a milk sample 10 days after treatment for a follicular cyst. If a white result (high progesterone) is obtained, the prescribed treatment was effective.

MILK RING TEST

In lactating animals, the MRT can be used for screening herds for brucellosis. In large herds (>100 lactating cows), the sensitivity of the test becomes less reliable. False-positive reactions may occur in recently vaccinated cattle (less than 4 months) or in samples containing abnormal milk, such as colostrum or that due to mastitis.

ANTIGEN PRODUCTION

MRT antigen is prepared from concentrated, killed B. abortus strain 99 or 1119-3 cell suspension, grown as described previously. It is centrifuged at, for example, 23,000 gfor 10 minutes at 4°C, followed by resuspension in haematoxylin-staining solution. Various satisfactory methods are in use; one example is as follows: 100 ml of 4% (w/v) haematoxylin (Cl No. 75290) dissolved in 95% ethanol is added to a solution of ammonium aluminium sulphate (5 g) in 100 ml of distilled water + 48 ml of glycerol. 2 ml of freshly prepared 10% (w/v) sodium iodate is added to the solution. After standing for 30 minutes at room temperature, the deep purple solution is added to 940 ml of 10% (w/v) ammonium aluminium sulphate in distilled water. The pH of this mixture is adjusted to 3.1, and the solution must be aged by storage at room temperature in the dark for 45- 90 days.

Before use, the staining solution is shaken and filtered through cotton wool. The packed cells are suspended in the staining solution at the rate of 1 g per 30 ml stain, and held at room temperature for 48 hours (some laboratories prefer to heat at 80°C for 10 minutes instead). The stained cells are then deposited by centrifugation, and washed three times in a solution of sodium chloride (6.4 g), 85% lactic acid (1.5 ml) and 10% sodium hydroxide (4.4 ml) in 1.6 litres of distilled water, final pH 3.0.

The washed cells are resuspended at the rate of 1 g in 27 ml of a diluent consisting of 0.5% phenol saline, adjusted to pH 4.0 by the addition of 0.1 M citric acid (approximately 2.5 ml) and 0.5 M disodium hydrogen phosphate (approximately 1 ml) and maintained at 4°C for 24 hours. The mixture is filtered through cotton wool, the pH is checked, and the PCV is determined and adjusted to approximately 4%. The sensitivity of the antigen is then repeatedly checked against that of a previously standardised batch using a panel of samples of varying degrees of reaction prepared by diluting a positive working standard serum (calibrated against the ISABS) in negative milk (7) The antigen must be stored at 4°C and not frozen. The pH of the antigen should be between 3.3 and 3.7 and its colour should be dark blue. There might be a little free stain in the supernatant of a centrifuged sample. When diluted in milk from a brucellosis-free animal, the antigen must produce a uniform coloration of the milk layer with no deposit and no coloration of the cream layer.

Test procedure

The test is performed by adding 30 μl of antigen to a 1 ml volume of whole milk that has been stored for at least 24 hours at 4°C. The height of the milk column in the tube must be at least 25 mm. If bulk tank samples from large herds are to be examined, the volume of milk should be increased to 3 ml. The milk samples must not have been frozen, heated or subjected to violent shaking. The milk/antigen mixtures are incubated at 37°C for 1 hour, together with positive and negative control samples. A strongly positive reaction is indicated by formation of a dark blue ring above a white milk column. Any blue layer at the interface of milk and cream should be considered positive as it might be significant, especially in large herds. The test is considered to be negative if the colour of the underlying milk remains homogeneously dispersed in the milk column. If the milk at the bottom of the tube becomes gradually whitened, the result is regarded as inconclusive and the test should be repeated.

MILK FAT TEST

Farmers and consumers alike are increasingly interested in the composition and nutritional value of milk. Consumers are bombarded daily with advertisements, articles and books that tell them what they should eat and drink and what they should not. No food product escapes this process and some of the media coverage given to dairy products has been less than positive.

Dairy farmers themselves try to influence consumers' actions and attitudes towards milk and milk products through advertising and education. Despite these efforts, farmers are beginning to realize that changes in composition may be necessary if dairy products are to maintain their share of the market. Naturally, much can be done to alter milk and its products at the processing level but this article will look at some possibilities for manipulating milk composition on the farm.

The first part of the article deals with the nutritional component of milk that has traditionally received the most attention - the fat content. Fat content may someday be considered a negative component in milk pricing and marketing, but for the immediate future, milk fat levels will continue to be important to New Brunswick producers. Milk fat level is also an item over which farmers can exert considerable control.

Breed Differences

Some major differences due to breed exist in milk composition and in milk fat levels in particular. Typical values for the common dairy breeds.

Table. Typical Breed Averages for Fat, Protein and Lactose Content of Milk.

Component	Holstein	Jersey	Ayrshire	Guernsey
Fat, %	3.6	4.8	3.9	4.6
Protein, %	3.2	3.8	3.3	3.6
Lactose, %	4.7	4.9	4.9	4.9

Within Breed Variation

In addition to the differences in milk composition between breeds, large differences exist between cows within the same breed. Fat content is the most variable component followed by protein content with lactose being the least variable. Since, the percentage of milk solids tends to go down as milk yield goes up, total yield of components (i.e. total solids or any of the individual components making up the solids) is often the more significant criteria than the percentage of a component.

In other words, yield of total solids or any of its individual components is the more important consideration. For this reason, milk yield data are often viewed on a fat-corrected basis, thus the term FCM or fat-corrected milk. The genetic potential of the cow sets the upper limit for milk fat test - manipulating feeding programmes or management practices will be to no avail if fat test is limited to a low level by genetics.

Non-nutritional and Non-genetic Factors

Factors other than nutrition and genetics can cause a change in milk fat test. Some of the more important factors are summarised.

Table. the Effect of Factors other than Nutrition and Genetics on Milk Fat Test

Stage of lactation	fat test decreases 1st two months of lactation, increases thereafter.
Cow age	fat test declines with age
Season	fat test usually lowest in summer, highest in winter.
Milking Interval	the milk taken following the shorter time period will have a higher test.
Mastitis	slight decrease in fat test with mastitis.

Milk Fat and Diet

Milk fat level is quite variable and reasonably responsive to diet. A large portion of the variation observed in milk fat can be attributed to changes in fatty acids produced in rumen fermentation. More precisely, the higher the ratio of acetate to propionate in the rumen, the higher the fat content of the milk will be.

Forage Level in the Diet

The most common dietary cause of a low fat test is a diet containing a low level of forage and a high level of concentrate. The low fat test observed in this situation is often confounded by the fact that low forage diets are often associated with high milk production and high milk production itself tends to lower milk fat test. Many of the rules of thumb applied to forage feeding are aimed at maintaining the type of rumen fermentation that will give a normal milk fat test. Commonly used guidelines are a minimum of35% forage in the diet dry matter or a daily forage dry matter intake equivalent to 1.5% of body weight.

A diet which maintains "normal" fat test without sacrificing production is usually a healthy one and often the most economic. For these reasons, fat test is often used as a barometer of the nutritional status of a herd. Since, fat test is so highly variable between cows one must be cautious in using individual cow data for ration evaluation - group or herd averages will always give a more accurate picture.

Fibre in the Diet

Although we often view forage as the dietary component required to maintain fat test, it is in fact the fibre in the forage that is crucial. Fibre, usually measured as ADF (acid detergent fibre) or NDF (neutral detergent fibre), is the dietary factor that most accurately indicates the likelihood of a diet maintaining fat test. Since, most feeds in a cow's diet contribute some fibre, the overall dietary fibre level needs to be considered, not just the contribution from forage. A guideline sometimes applied to dairy diets is a recommended minimum daily NDF intake equal to 1.1-1.2% of body weight.

Unfortunately, fibre level alone will not always predict a diet's ability to maintain milk fat test. An adequate length of cut of the forage (i.e. the length of the fibre) is also an important factor. Whether a forage is stored as finely chopped silage or as long hay will have no effect on the level of fibre as determined in a feed analysis lab. However, the forage stored as hay will be more effective in maintaining milk fat test. A large part of this difference comes about because the long forage (i.e. long fibre) stimulates cud-chewing and saliva flow. The high buffering capacity of cow saliva helps to maintain rumen pH and acetate levels and fat test.

A similar comparison can be made between a coarse grain mix and a similar pelleted one. Although a chemical analysis will say they are the same, the coarse feed will be a more effective source of fibre and will more likely maintain fat test.

Type of Grain

The type of grain fed also influences rumen fermentation and milk fat test. The starch in corn for example is quite slowly degraded in the rumen and tends to maintain higher and more uniform rumen pH, higher rumen acetate levels and higher milk fat test when compared to a more rapidly degraded starch such as barley. However, rumen fermentation is such a dynamic and complex process that factors such as dietary fibre level and feeding schedule often override the effects of grain source.

Dietary Protein Level

In most practical feeding situations, dietary protein level does not influence milk fat test. In some cases, however, researchers and farmers observe an improvement in fat test when more protein is fed. This is likely due to an improvement in rumen fermentation and fibre digestion with an accompanying increase in the proportion of acetate in the rumen resulting in a higher fat test.

On the other hand, in cases where extra protein causes an increase in milk volume, fat test percentage could drop simply due to the effect of dilution. Despite the general lack of effect of dietary protein on fat test, feeding the amino acid methionine (a protein constituent) often does improve fat test. Some commercial protein topdresses include supplemental methionine for this reason.

Dietary Fat

Traditionally, dairy cows received relatively small amounts of lipid (fats and oils) in their diets. In more recent years however, many high producing cows have been fed fat either in the form of oilseeds, such as soybeans, or as added fat. Data on the effect of dietary fat level on butterfat test can be confusing, with positive, negative and neutral effects all commonly recorded. The primary factors involved in these apparent discrepancies are the type of fat and level at which it is fed.

Unsaturated fats, (they usually exist as oils at room temperatures), are detrimental to rumen fermentation and fibre digestion and often depress feed intake. Feeding this type of fat, especially when it is not incorporated in a seed such as whole soybeans, often causes a depression in fat test. More saturated fats such as tallow however, do not have the same negative effects as unsaturated oils and, when fed in moderate amounts will often cause an increase in milk fat test.

Rumen escape or rumen inert fats such as the commercial product Megalac also have the potential for increasing fat test especially in 1st-calf heifers. Another aspect in the feeding of rumen inert fats is how they change the composition of the milk fat, rather than the milk fat level. A significant portion of 'rumen by-pass' fat may be transferred unaltered into the milk, resulting in a change in the milk fat composition, especially the balance of saturated to unsaturated fat. This could have important implications as a way to alter milk fat composition to meet specific market requirements.

Buffers

Buffers such as sodium bicarbonate and magnesium oxide have been used as feed additives in dairy diets for years. They can be effective in maintaining proper rumen pH and alleviating fat depression. They will not improve fat test unless the test has been depressed due to excessive acidity in the rumen. As with other strategies, buffers will not improve a "normal" or a genetically depressed test.

Feeding Schedule and Sequence

Some low-milk fat problems can be alleviated with changes in feeding schedules, feeding sequences or feeding system. For example, TMR (total mixed ration) feeding has been promoted and used as a method of minimizing the pH-depressing effects of high levels of grain feeding. Buffering the rumen by feeding forage prior to grain is another widely promoted practice that is aimed at stabilizing rumen pH and maintaining fat test. Although research on multiple feedings has been limited, review articles on work-to-date indicate that an increase in milk fat is the most likely benefit to be obtained from multiple feedings.

Niacin

Niacin is a B vitamin that is important in fat metabolism. Fat cows often show a milk production response to niacin supplementation in early lactation. Although research results are not conclusive or dramatic, this increased production is often accompanied by an increase in fat test.

5

Processing Techniques of Milk and Milk Products

Milk can be stored longer if it has been processed. Extended storage is possible if you are able to control the growth of micro-organisms. The processing technique used will determine the storage life of the milk and dairy products.

The following rules should be followed during the production, storage and processing of milk:

- Always wash your hands and avoid putting them in the milk if not necessary.
- All equipment used during processing is properly cleaned and disinfected if needed.
- Take care that no dirt particles or insects enter into the milk.
- Try to prevent the use of copper utensils. (Copper can give off flavours in butter and milk.)
- Do not expose milk to sunlight; store it in a dark place.
- The use of a thermometer is recommended.
- Make sure that milk used for consumption has always been boiled or pasteurised.
- Never store raw (*i.e.* unheated) milk if it is not immediately cooled below 4°C.
- Never drink raw milk because it may contain pathogenic bacteria like tubercular bacteria and salmonella.

The following processing techniques will be dealt with in this chapter:

- Pasteurisation
- Cooling
- Souring (acidification)
- Creaming.

Heating and cooling are in fact ways to preserve milk, but for convenience as suggested, deal with them under the heading of processing techniques.

HYGIENE DURING MILKING

There are several possible causes of contamination during milking. In a normal, healthy cow very low numbers of bacteria are found inside the udder and the teats. Cows possess various mechanisms to prevent the entry of bacteria. To avoid problems while milking, it is important that an animal become accustomed to the activity. It will then know that it will be milked, and will react positively to it. Such positive behaviour can start if, for instance, it hears milk cans clanging, feels its udder being cleaned, etc. Then the animal is easier to milk and gives more milk.

Stress and unrest make the cows move too much and kick; consequently more dirt and manure can enter into the milk. When a cow has an udder infection (mastitis), its milk will be contaminated with the bacteria that cause the udder infection, and that may produce pus and sometimes blood. Milk from these animals should not be used in any way. Mastitis can be prevented by maintaining good hygiene and avoiding injury to the teats during milking.

An infected udder is not always easy to see. When an udder infection occurs, it is advisable to remove milk from the udder very frequently (*e.g.* every 3 hours by hand). The number of microorganisms in the udder is thus reduced. Be aware, however, that milking an infected udder by machine or by hand is often painful for the animal. The animal will kick frequently and this can be an important source of contamination of healthy cows. Bacteria can be transferred from the skin or teats to the milk, even with healthy dairy cattle.

It is therefore important to clean the udder before milking. Wipe the udder clean with a dry, clean, preferably disposable cloth to prevent infection. If the teats or udder are really dirty, they must first be washed with clean, hand-warm water and a clean cloth and then dried with a clean towel. Cleaning the udder improves the cleanliness of the milk and makes milking easier. Skin and hair can also be sources of infection. Do not feed animals first before milking, it may create a lot of dust. The floor is clean, and be careful when clearing dung, mud or dust.

A clean, well-illuminated milking place and fresh surrounding air are essential to maintaining good hygiene. Insects such as flies and cockroaches can also be sources of infection. Try to control them as they can carry many bacteria and viruses. When milking, the milk is caught in a pail or bucket. Dirty milking equipment is the main source of infection of milk. If residues of milk remain in the equipment because of improper cleaning and drying, bacteria will develop in these residues. These bacteria are already accustomed to the milk and will multiply rather quickly during transport and storage of milk in the equipment.

Use pails and buckets that are smooth on the inside, for instance seamless metal buckets. All milking equipment should be thoroughly cleaned immediately after each use. Use soap or other detergents if necessary. Make sure that the water used is clean. If you are in doubt, boil it for several minutes or add chlorine. Very important: after cleaning, the equipment should be stored upside down in such a way that the inside of the buckets and cans dry. This prevents the remaining bacteria from growing.

The person milking plays the most important part in maintaining proper hygiene during production. He or she keeps an eye on the condition of the animal, chooses the milking place and cleans all the equipment. He or she should have clean hands and wear clean clothes. If the milker suffers from tuberculosis, salmonella infection, dysentery or some other disease, the risk of contamination of the milk becomes very high; it would be wise to have somebody else take over. This is also the case if the milker has open wounds or ulcers.

Hygiene during Storage and Processing

By now you should know that milk should be processed as quickly as possible after milking and that it should be properly stored in order to minimise its chances of spoiling. It is best to filter fresh milk through a filter or clean cloth. This will remove visible dirt that might have entered into the milk. Clean or replace the cloth during filtering or filter the milk several times. The cloth should be thoroughly cleaned after use and then left to dry in the sun.

In tropical conditions, raw milk, *i.e.* non-pasteurised milk, goes off within a few hours. It must therefore be kept cool and quickly pasteurised and again cooled to a temperature of 4°C if possible. Properly pasteurised and cooled milk can be kept for a few days, even in a warm climate. If you are not able to cool milk below 10°C, then do not mix different batches.

Even if the older milk is still good, you will end up with an increase in bacterial growth and reduction of the overall quality. Use clean equipment for storage. Containers that are clear, such as glass, should be stored in the dark as light reduces the quality of milk. Clean your equipment with clean water.

Cleaning and Disinfection

Utensils must be cleaned in such a way that all dirt, food residues, feed and micro-organisms are removed from the surface of the equipment. Dirty saucepans, jugs, milking equipment and utensils should be cleaned immediately after use. Washing soda (sodium carbonate) dissolved in hot water is an excellent cleaning agent. It may be useful to disinfect equipment in order to kill any remaining harmful micro-organisms. You can use a chloride solution such as bleach (sodium hypochlorite).

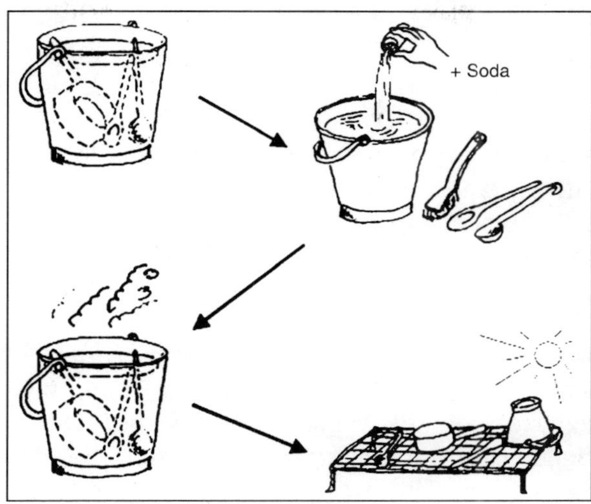

Fig. Cleaning Utensils

A proper way of cleaning your equipment is the following:

- Start cleaning immediately after milking, so that milk residues will not dry and stick on the buckets and utensils.
- Rinse well with water.
- Scrub the tools in a hot soda solution (1.5 tablespoons of soda to 5 litres of water), using a small amount of water to dissolve the soda before adding it to the rest of the water.
- Rinse well with hot water.
- Buckets, tubs, etc., should be turned upside down on a rack during storage; the water can then drain and no dirt or dust can enter. Let the utensils dry to prevent bacterial growth.

Well-cleaned tools are nearly sterile, only a small part of the bacteria remains on the tools. If these tools dry during storage hardly any bacteria will be present. In that case disinfection is not necessary. Tools which are used for storage of pasteurised milk or for cheese making and which do not get a heat treatment together with the milk can be disinfected after cleaning or before use.

Proceed as follows:

- Clean all your equipment properly. The following step will be ineffective if the utensils are not clean to start with.
- Disinfect in a chloride or bleach solution after cleaning or shortly before use. Add 2 tablespoons of bleach per 4. 5 litres of water.

It is advisable to use stainless steel equipment, cheesecloth and wooden utensils. Tools or any other equipment made from aluminium should not be washed in a strong soda solution, as soda attacks aluminium. Iron utensils will rust in a strong chloride solution. Therefore rinse and dry these utensils immediately after cleaning and disinfection.

If you have no cleaning agents-like soda-or disinfectants, you can disinfect your equipment as follows:

- Thoroughly clean the utensils using clean water.
- Rinse with a soap solution.
- Dry the equipment on a rack in the sun upside down or rinse with boiling water.

PASTEURISATION OF MILK

As you by now know, milk contains certain micro-organisms that can spoil it. These bacteria grow best at temperatures between 10°C and 40°C. It is therefore important to cool milk as quickly as possible. This can be difficult in the tropics if no cold water or refrigerators are available. Most bacteria will be destroyed during heating. The most effective temperature depends on the heating time. In other words, heating for a longer period at a lower temperature can be as effective as heating for a shorter period but at a higher temperature.

Pasteurisation improves the safety and storage life of a product, while the taste hardly changes and the loss of vitamins is minimal. A distinction is made between low and high pasteurisation. Although high pasteurisation initially kills more bacteria, the resulting milk can usually not be kept as long, because the high pasteurisation temperature stimulates spores of some bacteria to germinate. Moreover, the taste of high pasteurised milk has more or less the flavour of boiled milk. Pasteurised milk can be kept for about one week at 4-6°C if no re-infection takes place.

The pasteurisation temperature to be used depends on the product to be made of the milk:

- Low pasteurisation is used for milk for direct consumption and cheese.
- High pasteurisation is used for yoghurt, butter and kefir.

Table. Time–temperature Combinations for Pasteurising Milk

Remarks	Time	Temperature	
Low Pasteurising Quantities>5 litres	30 minutes	63°C	
	3 minutes	68°C	Small
quantities			
	20 seconds	72°C*)	
Industrial equipment			
High Pasteurising	2 minutes	82°C	
	20 seconds	85°C*)	

Notes:

*) Continuous flow system; not for small-scale processing

If there is no thermometer to measure the exact temperature, heat the milk to its boiling point.

Pasteurisation Methods

The following method A is suitable if you are able to accurately control both temperature and time. Method B is more hygienic, if the exact temperature of the milk is unknown.

You will need:

- Raw milk, a heat source, a saucepan with a thick bottom that is smooth on the inside, a thermometer and a means of cooling the milk that has been heated
- For method A: a clean wooden spoon
- For method B: glass jars with lids or bottles with tops, or plastic bags and sealing equipment.

Method A

Put the milk in a clean saucepan and heat it to 68°C, stirring continuously. Keep the milk at that temperature for at least 3 minutes.

Method B

Clean the glass jars with lids or bottles with tops. Fill them with milk and close the lids. Submerge the jars and bottles in a large pan filled with water. Heat everything to 80°C and keep it at that temperature for at least 10 minutes. When using method A, let the milk cool down as quickly as possible. The best storage temperature is 4°C.

If you are using method B but are not able to maintain a constant temperature of 80°C, the best alternative is to heat the water in the pan until it boils and keep it boiling for some time. It is important to store the pasteurised or boiled milk at 4°C. At this temperature it can be kept up to one week. Handle the boiled or pasteurised milk with care to prevent re-infection.

WHAT IS PASTEURIZATION?

In 1864, a French man named Louis Pasteur discovered that liquids such as milk could be heated to a temperature slightly below boiling and held there for a set amount of time to eliminate the most harmful bacteria. The process of pasteurization is named after Louis Pasteur in recognition of his immense contribution tofood safety and disease theory. Grocery stores carry a wide array of pasteurized goods including milk, juices, non-dairy milks, and other similar food products. Many nations require that foods be pasteurized for safety. Pasteurization relies on the principle that most harmful bacterial can be killed by heat.

The most effective way to kill bacteria is boiling, but this compromises the flavour of the liquid. Pasteurization strikes a happy medium, keeping the flavour delicious while making the food safer. In addition to minimizing the risk of sickness, pasteurization also makes foods more shelf stable and less likely to rot, meaning that fresh dairy products and juices are available to more people.

There are two primary methods of pasteurization: the liquid can be heated to 145 degrees Fahrenheit (63 degrees Celsius) and held there for at least thirty minutes, or the liquid can be flash pasteurized at 161 degrees Fahrenheit (72 degrees Celsius) for a minimum of 16 seconds. Pasteurization can be done using a continuous method, where the liquid flows through a pasteurization system, or by using a batch method, where one batch of the liquid is pasteurized at a time. Continuous pasteurization is popular for large producers, because it does not slow the supply line as much as batch pasteurization does. Pasteurization must be performed on clean equipment. If bacteria is introduced after the liquid has been pasteurized, it can colonize it and potentially cause an outbreak of food borne illness. For this reason, companies which perform pasteurization are subject to frequent inspections to ensure that the equipment they are using is safe, and that the liquids they are pasteurizing are being handled correctly. After pasteurization, bacteria can still appear. It is important that foods be safely handled and stored at every step of the supply process from animal, fruit or vegetable to stomach. In most cases, after food is pasteurized it should be refrigerated. The food is held in refrigeration until it is shipped out in cooled trucks to grocery stores, which store the food under refrigeration until consumers purchase it. Home consumers are responsible for following directives regarding temperature to ensure that the foods they are consuming are safe.

The process of pasteurization was named after Louis Pasteur who discovered that spoilage organisms could be inactivated in wine by applying heat at temperatures below its boiling point. The process was later applied to milk and remains the most important operation in the processing of milk. The heating of every particle of milk or milk product to a specific temperature for a specified period of time without allowing recontamination of that milk or milk product during the heat treatment process.

*Purpose Ther*How do I disinfect my pr*e are two distinct purposes for the process of milk pasteurization*:

1. *Public Health Aspect*: To make milk and milk products safe for human consumption by destroying all bacteria that may be harmful to health (pathogens)
2. *Keeping Quality Aspect:* To improve the keeping quality of milk and milk products. Pasteurization can destroy some undesirable enzymes and many spoilage bacteria. Shelf life can be 7, 10, 14 or up to 16 days.

The extent of microorganism inactivation depends on the combination of temperature and holding time. Minimum temperature and time requirements for milk pasteurization are based on thermal death time studies for the most heat resistant pathogen found in milk, Coxelliae burnettii. Thermal lethality determinationsrequire the applications of microbiology to appropriate processing determinations. An overview can be found here. To ensure destruction of all pathogenic microorganisms, time and temperature combinations of the pasteurization process are highly regulated:

ONTARIO PASTEURIZATION REGULATIONS

Milk

63° C for not less than 30 min.,

72° C for not less than 16 sec.,

Or equivalent destruction of pathogens and the enzyme phosphatase as permitted by Ontario Provincial Government authorities. Milk is deemed pasteurized if it tests negative for alkaline phosphatase.

Frozen dairy dessert mix (ice cream or ice milk, egg nog):

- At least 69° C for not less than 30 min;
- At least 80° C for not less than 25 sec;

other time temperature combinations must be approved (e.g. 83° C/16 sec). Milk based products- with 10% mf or higher, or added sugar (cream, chocolate milk, etc) 66° C/30 min, 75° C/16 sec. There has also been some progress with low temperature pasteurization methods using membrane processing technology..

METHODS OF PASTEURIZATION

There are two basic methods, batch or continuous.

Batch Method

The batch method uses a vat pasteurizer which consists of a jacketed vat surrounded by either circulating water, steam or heating coils of water or steam.

Fig Batch Pasteurizer (26 KB)

In the vat the milk is heated and held throughout the holding period while being agitated. The milk may be cooled in the vat or removed hot after the holding time is completed for every particle. As a modification, the milk may be partially heated in tubular or plate heater before entering the vat.

This method has very little use for milk but some use for milk by-products (e.g. creams, chocolate) and special batches. The vat is used extensivly in the ice cream industry for mix quality reasons other than microbial reasons.

CONTINUOUS METHOD

Continuous process method has several advantages over the vat method, the most important being time and energy saving. For most continuous processing, a high temperature short time (HTST) pasteurizer is used. The heat treatment is accomplished using a plate heat exchanger. This piece of equipment consists of a stack of corrugated stainless steel plates clamped together in a frame. There are several flow patterns that can be used. Gaskets are used to define the boundaries of the channels and to prevent leakage. The heating medium can be vacuum steam or hot water.

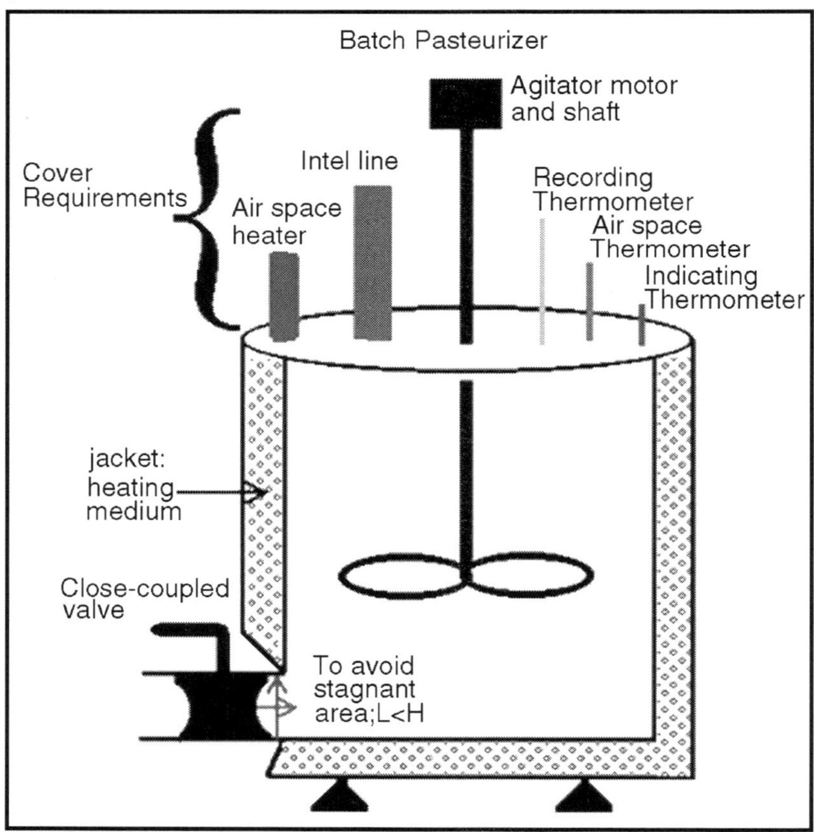

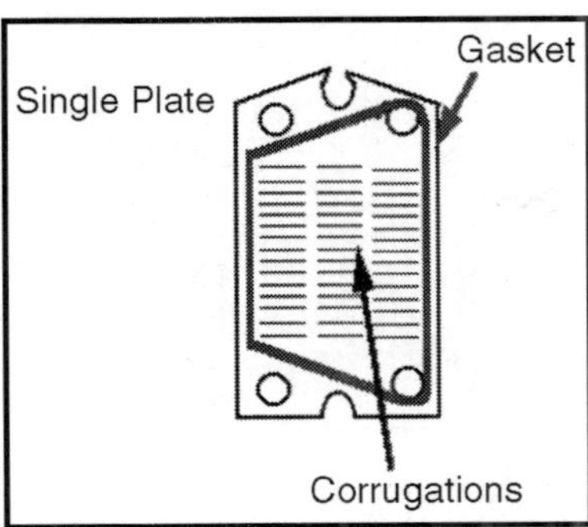

Fig. Plate Heat Exchanger 26 KB

HTST Milk Flow Overview

This overview is meant as an introduction and a summary. Each piece of HTST equipment will be discussed in further detail later. Cold raw milk at 4° C in a constant level tank is drawn into the regenerator section of pasteurizer. Here it is warmed to approximately 57° C - 68° C by heat given up by hot pasteurized milk flowing in a counter current direction on the opposite side of thin, stainless steel plates. The raw milk, still under suction, passes through a positive displacement timing pump which delivers it under positive pressure through the rest of the HTST system.

The raw milk is forced through the heater section where hot water on opposite sides of the plates heat milk to a temperature of at least 72° C. The milk, at pasteurization temperature and under pressure, flows through the holding tube where it is held for at least 16 sec. The maximum velocity is governed by the speed of the timing pump, diametre and length of the holding tube, and surface friction. After passing temperature sensors of an indicating thermometre and a recorder-controller at the end of the holding tube, milk passes into the flow diversion device (FDD).

The FDD assumes a forward-flow position if the milk passes the recorder-controller at the preset cut-in temperature (>72° C). The FDD remains in normal position which is in diverted-flow if milk has not achieved preset cut-in temperature. The improperly heated milk flows through the diverted flow line of the FDD back to the raw milk constant level tank.

Properly heated milk flows through the forward flow part of the FDD to the pasteurized milk regenerator section where it gives up heat to the raw product and in turn is cooled to approximately 32° C - 9° C. The warm milk passes through the cooling section where it is cooled to 4° C or below by coolant on the opposite sides of the thin, stainless steel plates. The cold, pasteurized milk passes through a vacuum breaker at least 12 inches above the highest raw milk in the HTST system then on to a storage tank filler for packaging.

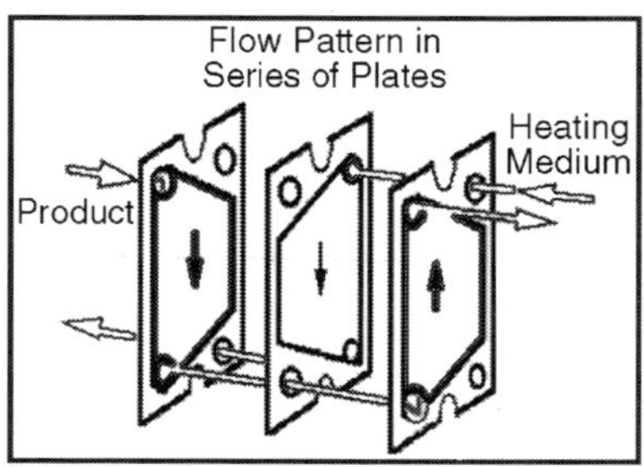

Fig. Basic Flow - HTST Pasteurization 17 KB

Holding Time

When fluids move through a pipe, either of two distinct types of flow can be observed. The first is known as turbulent flow which occurs at high velocity and in which eddies are present moving in all directions and at all angles to the normal line of flow. The second type is streamline, or laminar flow which occurs at low velocities and shows no eddy currents.

The Reynolds number, is used to predict whether laminar or turbulent flow will exist in a pipe:

Re < 2100 laminar

Re > 4000 fully developed turbulent flow

There is an impact of these flow patterns on holding time calculations and the assessment of proper holding tube lengths. The holding time is determined by timing the interval for an added trace substance (salt) to pass through the holder. The time interval of the fastest particle of milk is desired. Thus the results found with water are converted to the milk flow time by formulation since a pump may not deliver the same amount of milk as it does water.

Note: the formulation assumes flow patterns are the same for milk and water. If they are not, how would this affect the efficiency of the pasteurization process?

PRESSURE DIFFERENTIAL

For contiunuous pasteurizing, it is important to maintain a higher pressure on the pasteurized side of the heat exchanger. By keeping the pasteurized milk at least 1 psi higher than raw milk in regenerator, it prevents contamination of pasteurized milk with raw milk in event that a pin-hole leak develops in thin stainless steel plates.

Thispressure differential is maintained using a timing pump in simple systems, and differential pressure controllers and back pressure flow regulators at the chilled pasteurization outlet in more complex systems. The position of the timing pump is crucial so that there is suction on the raw regenerator side and pushes milk under pressure through pasteurized regenerator.

There are several other factors involved in maintaining the pressure differential:

- The balance tank overflow level must be less than the level of lowest milk passage in the regenerator
- Properly installed booster pump is all that is permitted between balance tank and raw regenerator
- No pump after pasteurized milk outlet to vacuum breaker
- There must be greater than a 12 inch vertical rise to the vacuum breaker
- The raw regenerator drains freely to balance tank at shut-down

BASIC COMPONENT EQUIPMENT OF HTST PASTEURIZER

Balance Tank

The balance, or constant level tank provides a constant supply of milk. It is equipped with a float valve assembly which controls the liquid level nearly constant ensuring uniform head pressure on the product leaving the tank. The overflow level must always be below the level of lowest milk passage in regenerator. It, therefore, helps to maintain a higher pressure on the pasteurized side of the heat exchanger. The balance tank also prevents air from entering the pasteurizer by placing the top of the outlet pipe lower than the lowest point in the tank and creating downward slopes of at least 2%. The balance tank provides a means for recirculation of diverted or pasteurized milk.

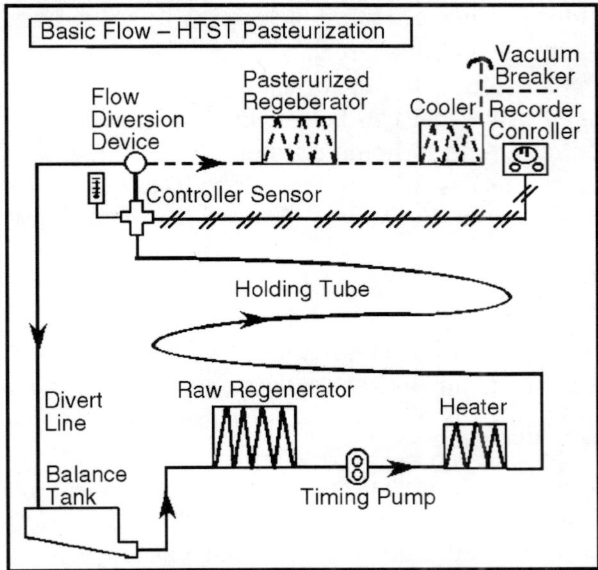

Fig. Balance Tank 17 KB

Regenerator

Heating and cooling energy can be saved by using a regenerator which utilizes the heat content of the pasteurized milk to warm the incoming cold milk.

Its efficiency may be calculated as follows: % regeneration = temp. increase due to regenerator/total temp. increase

For example: Cold milk entering system at 4° C, after regeneration at 65° C, and final temperature of 72° C would have an 89.7% regeneration:

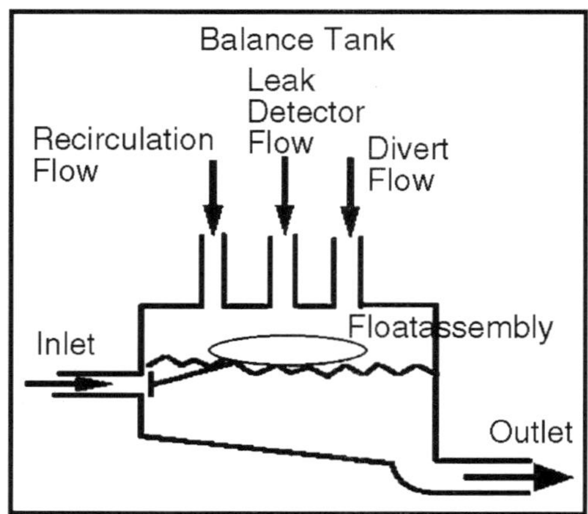

Timing Pump

The timing pump draws product through the raw regenerator and pushes milk under pressure through pasteurized regenerator. It governs the rate of flow through the holding tube. It must be a positive displacement pump equipped with variable speed drive that can be legally sealed at the maximum rate to give minimum holding time in holding tubes. It also must be interwired so it only operates when FDD is fully forward or fully diverted, and must be"fail-safe". A centrifigal pump with magnetic flow metre and controller may also be used.

Holding Tube

Must slope upwards 1/4"/ft. in direction of flow to eliminate air entrapment so nothing flows faster at air pocket restrictions.

Indicating Thermometre

The indicating thermometre is considered the most accurate temperature measurement. It is the official temperature to which the safety thermal limit recorder (STLR) is adjusted. The probe should sit as close as possible to STLR probe and be located not greater than 18 inches upstream of the flow diversion device.

Recorder-Controller (STLR)

The STLR records the temperature of the milk and the time of day. It monitors, controls and records the position of the flow diversion device (FDD) and supplies power to the FDD during forward flow. There are both pneumatic and electronic types of controllers. The operator is responsible for recording the date, shift, equipment, ID, product and amount, indicating thermometre temperature, cleansing cycles, cut in and cut out temperatures, any connects for unusual circumstances, and his/her signature.

Flow Diversion Device (FDD)

Also called the flow diversion valve (FDV), it is located at the downstream end of the upward sloping holding tube. It is essentially a 3-way valve, which, at temperatures greater than 72° C,opens to forward flow. This step requires power. At temperatures less than 72° C, the valve recloses to the normal position and diverts the milk back to the balance tank. It is important to note that the FDD operates on the measured temperature, not time, at the end of the holding period.

There are two types of FDD:

1. *Single stem*: An older valve system that has the disadvantage that it can't be cleaned in place.
2. *Dual stem*: Consists of 2 valves in series for additional fail safe systems. This FDD can be cleaned in place and is more suited for automation.

$$\frac{65\text{-}4}{72\text{-}4}\ 89.7$$

Fig. Flow Diversion Devices 17 KB

Vacuum Breaker

At the pasteurized product discharge is a vacuum breaker which breaks to atmospheric pressure. It must be located greater than 12 inches above the highest point of raw product in system. It ensures that nothing downstream is creating suction on the pasteurized side.

Auxiliary Equipment

Booster Pump

It is centrifugal"stuffing" pump which supplies raw milk to the raw regenerator for the balance tank. It must be used in conjunction with pressure differential controlling device and shall operate only when timing pump is operating, proper pressures are achieved in regenerator, and system is in forward flow.

Homogenizer

The homogenizer may be used as timing pump. It is a positive pressure pump; if not, then it cannot supplement flow. Free circulation from outlet to inlet is required and the speed of the homogenizer must be greater than the rate of flow of the timing pump.

Magnetic Flow Metre and Centrifugal Pump Arrangements

Magnetic flow metres can be used to measure the flow rate. It is essentially a short piece of tubing (approximately 25 cm long) surrounded by a housing, inside of which are located coils that generate a magnetic field. When milk passes through the magnetic field, it causes a voltage to be induced, and the generated signal is directly proportional to velcoity. Application of the magnetic flow metre in the dairy industry has centred around its replacing the positive displacement timing pump as the metreing device in HTST pasteurizing systems, where with certain products the timing pump rotors reportedly wear out in a relatvely short period of time.

In operation, the electrical signal is sent by the magnetic flow metre to the flow controller, which determines what the actual flow is compared to the flow rate set by the operator. Since the magnetic flow metre continuously senses flow rate, it will signal the electronic controller if the actual flow exceeds the set flow rate for any reason.

If the flow rate is exceeded for any reason, the flow diversion device is put into diverted flow. A significant difference from the normal HTST system (with timing pump) comes into focus at this point. This system can be operated at a flow rate greater than (residence time less than) the legal limit. However, it will be in diverted flow and never in forward flow. Another magnetic flow metre based system with an AC variable frequency motor control drive on a centrifugal pump is also possible in lieu of a positive displacement metreing pump on a HTST pasteurizer.

This system does not use a control valve but rather the signal from the magnetic flow metre is transmitted to the AC variable frequency control to vary the speed of the centrifugal pump. The pump, then controls the flow rate of product through the system and its holding time in the holding tube.

Automated Public Health Controllers These systems are used for time and temperature control of HTST systems. There are concerns that with sequential control, the critical control points (CCP's) are not monitored all the time; if during the sequence it got held up, the CCP's would not be monitored. With operator control, changes can be made to the programme which might affect CCP's; the system is not easily sealed. No computer programme can be written completely error free in large systems; as complexity increases, so too do errors.

This gives rise to a need for specific regulations or computer controlled CCP's of public health significance:

- Dedicated computer - no other assignments, monitor all CCP's at least once/sec
- Not under control of any other computer system or override system, *i.e.,* network
- Separate computer on each pasteurizer
- I/O bus for outputs only, to other computers no inputs from other computers
- On loss of power - public health computers should revert to fail safe position (e.g. divert)
- Last state switches during power up must be fail safe position
- Programmes in ROM - tapes/disks not acceptable
- Inputs must be sealed, modem must be sealed, programme sealed
- No operator override switches
- Proper calibration procedure during that printing - Public health computer must not leave public health control for > 1 sec and upon return must complete 1 full cycle before returning to printing
- FDV position must be monitored and temperature in holding tube recorded during change in FDV position
- Download from ROM to RAM upon startup
- Integrated with CIP computer which can be programmed *e.g.,* FDV, booster pump controllable by CIP computer when in CIP made only

PASTEURISED MILK PRODUCTS

SALMONELLA

Salmonellae are not able to survive the typical minimum pasteurisation processes generally prescribed in legislation. Therefore, their presence indicates that the process has not been carried out effectively, or that post-process contamination has occurred. For example, an outbreak of salmonellosis in Kentucky in 1984 was associated with pasteurised milk, but an investigation of the dairy concerned showed that pasteurisation temperatures were inadequate, and could have been as low as 54.5°C for 30 minutes.

An outbreak caused by *Salmonella braenderup* in the UK in 1986 was also associated with pasteurised milk, and on this occasion the pasteuriser was found to be poorly designed and incorrectly operated, probably resulting in the application of an inadequate heat treatment. In 1985, one of the largest outbreaks of salmonellosis in US history occurred in Illinois. Almost 200,000 people were affected, and were associated with pasteurised low-fat 2% milk contaminated with *S. typhimurium*. Investigations at the dairy plant involved revealed no evidence of inadequate pasteurisation, and the outbreak strain was not found to be abnormally heat resistant. Although the cause of the outbreak has never been completely explained, the investigation did discover a possible cross-connection between raw and pasteurised milk, which may have been the source of contamination. In 1998, an outbreak of salmonellosis in Lancashire, caused by a multiresistant strain of *S. typhimurium* DT104, affected 86 people. This outbreak was also linked to defective pasteurisation of milk at a dairy on a local farm.

Consumption of raw milk or raw milk products have been responsible for 62 and 29 cases of diarrheal illness caused by *S. typhimurium* in 2003 and 2007, respectively, in the states of Ohio and Pennsylvania. Since salmonellae are occasional contaminants of raw milk, they may sometimes enter the processing environment. It is very important that contamination of the post-pasteurisation plant is not allowed to occur and effective precautions and monitoring procedures, based on HACCP principles, are necessary to prevent this.

Campylobacter Spp.

Campylobacter spp. are not capable of surviving milk pasteurisation treatments, and cannot grow in raw or pasteurised milk, although they are able to survive for long periods in milk at refrigeration temperatures. Nonetheless, outbreaks of campylo-bacteriosis associated with pasteurised milk have occurred. For example, a large outbreak in the UK in 1979 caused by *Campylobacter jejuni* was estimated to have affected at least 2,500 schoolchildren, and was associated with free milk provided in schools. Although conclusive evidence was absent, it seems likely that raw milk may have bypassed the pasteurisation process.

A more recent outbreak in 2001 involved 75 people and was linked to the consumption of unpasteurised milk procured thorough a cow leasing programme. Birds are known to be an important reservoir of *Campylobacter* infection, and the tendency of some birds to peck through the foil tops of doorstep-delivered milk bottles is becoming recognised as an important source of infection in parts of the UK.

Some individual cases have been attributed to this cause, and, in one instance in 1990, the organism was isolated from the beaks of jackdaws and magpies as well as the contaminated milk. More recently, an outbreak thought to be associated with bird-pecked milk was reported.

Listeria Monocytogenes

There has been some discussion regarding the potential for *L. monocytogenes* in milk to survive pasteurisation. An outbreak of listeriosis in Massachusetts during 1983 resulted in 49 cases, 14 of whom subsequently died. Epidemiological evidence strongly suggested an association with consumption of pasteurised whole and low-fat milk, although this could not be confirmed microbiologically. The investigation failed to reveal any evidence of inadequate pasteurisation and the organism could not be found in environmental samples in the dairy, suggesting that postprocess contamination was unlikely.

However, samples of raw milk taken from farms supplying the plant were found to be positive for *L. monocytogenes* serotype 4b, and the investigators concluded that survival of some organisms through pasteurisation was the most likely cause of the outbreak. Three deaths and a miscarriage in Boston, USA between 2007-8 have been linked to presence of *Listeria* in pasteurised milk. So far, investigations have found nothing wrong with its pasteurisation process.

Furthermore, in a survey of pasteurised milk conducted in Spain, *L. monocytogenes* was recovered from six out of 28 samples heated at 78°C for 15 seconds. The explanation offered for both these findings was that the organisms might have been protected during heat treatment within leucocytes in the milk. However, this effect has not been conclusively demonstrated, and *L. monocytogenes* has not yet been shown to have survived pasteurisation in milk subjected to minimum HTST pasteurisation requirements of 71.7°C for 15 seconds.

For these reasons, it is currently accepted that existing pasteurisation processes are adequate to inactivate the organism in milk. *L. monocytogenes* is likely to be present in wet dairy processing environments, and post-process contamination is therefore a particular hazard. The organism has been shown to be capable of significantly more rapid growth in pasteurised milk than in raw milk at 7°C, and is also capable of growth at 4°C in pasteurised milk. Therefore, effective HACCP-based controls to prevent post-process contamination are critical, particularly the cleaning and sanitising of all milkcontact surfaces. Adequate temperature control is also important.

Verotoxigenic Escherichia Coli

Dairy cattle are an important reservoir for *E. coli* O157: H7 and this organism may therefore be present in raw milk, usually through faecal contamination. For this reason, raw milk is a high-risk food for this serious intestinal pathogen, and there have been a number of small outbreaks of infection associated with its consumption. However, *E. coli* O157: H7 is not a heat-resistant organism and there is no evidence that it is able to survive pasteurisation. Despite this, there have been outbreaks associated with pasteurised milk.

In 1994, an outbreak in Scotland affected over 100 people and was associated with consumption of pasteurised milk from a local dairy. The outbreak strain was eventually recovered from cows on one of the farms supplying the dairy, from a bulk milk tanker, and from a pipe transferring milk from the pasteuriser to the bottling machine. Whether this outbreak was the result of faulty pasteurisation or post-process contamination was unclear, but, in either case, the raw milk is likely to have been the original source of the organism. In 1999, a serious outbreak occurred in Cumbria in the north-west of England, which was also associated with pasteurised milk from a local dairy. There were at least 60 confirmed cases involved, and the cause was thought to be a fault in the operation of the pasteuriser. The first general outbreak of verocytotoxin-producing *E. coli* in Denmark occurred in 2004 and involved 25 patients; 18 children and seven adults. It was thought to be due to the consumption of a particular kind of organic milk from a small dairy.

Environmental and microbiological investigations at the suspected dairy did not confirm the presence of the outbreak strain, but the outbreak stopped once the dairy was closed and thoroughly cleaned. *E. coli* O157 is not reported to be able to grow in raw or pasteurised milk stored at 5°C, but may grow slowly at higher temperatures. However, since the infective dose of this pathogen is thought to be very low, effective pasteurisation and the prevention of post-process contamination are critical to ensure product safety.

Yersinia Enterocolitica

Although there has been a question about the ability of *Y. enterocolitica* to survive milk pasteurisation, the majority of the evidence indicates that it is inactivated. Three different strains of *Y. enterocolitica* were reported to have D-values of 0.24-0.96 minutes at 62.8°C. Therefore, the presence of the organism in pasteurised milk is likely to be the result of post-process contamination. There have been several *Y. enterocolitica* outbreaks associated with pasteurised milk.

In 1976, an outbreak affecting 36 children was associated with the consumption of contaminated chocolate milk. It was thought that the organism was introduced to the product during mixing of chocolate syrup with pasteurised milk, without any subsequent heat process. Another outbreak in 1982 was the largest foodborne yersiniosis outbreak ever recorded in the USA, and was also associated with pasteurised milk.

It is thought that several thousand people may have developed illness, although the organism was not isolated from milk or environmental samples at the dairy. It was found that surplus milk was used to feed pigs and that the crates used to transport this milk were stored on the ground at the farm and could have become contaminated with pig faeces.

Since pigs are a well known reservoir for *Y. enterocolitica*, it was thought that inadequate washing of the crates allowed the organism to survive in mud on them, and subsequently contaminate the external surfaces of milk cartons. *Y. enterocolitica* is capable of psychrotrophic growth, and could therefore multiply in pasteurised milk during storage. Measures should therefore be taken to prevent post-process contamination as with *L. monocytogenes.*

Staphylococcus Aureus

Staph. aureus is only rarely involved in food poisoning associated with consumption of pasteurised milk, although enterotoxigenic strains can be found as contaminants in raw milk. This may be because *Staph. aureus* does not generally grow at temperatures below 7°C, and enterotoxin production is inhibited at low temperatures. The organism is also known to be inhibited by the presence of competing species. Nevertheless, an outbreak in California affecting 500 school children was associated with chocolate-flavoured milk. The cause was thought to be growth of *Staph. aureus* in raw milk, and the subsequent persistence of the heat-stable enterotoxin through pasteurisation.

In June and July 2000, a very large outbreak of staphylococcal food poisoning was reported in Japan, associated with consumption of pasteurised low fat milk. Over 14,500 people were said to have been affected. The outbreak was unusual in that the thermal processes had destroyed staphylococci in milk but *Staphylococcus* enterotoxin A had retained enough activity to cause intoxication. SEA exposed at least twice to pasteurisation at 130°C for 4 or 2s retained both immunological and biological activities, although it had been partially inactivated.

Bacillus Spp

Psychrotrophic *Bacillus* spp. present in raw milk may survive pasteurisation and then become dominant in the pasteurised milk, potentially causing spoilage. Concerns have been expressed that some psychrotrophic strains of *B. cereus* may be able to produce toxin in milk at refrigeration temperatures, but it seems likely that obvious spoilage would occur before sufficient toxin production had taken place to cause illness. Even so, *B. cereus* was isolated at levels of 4x105/ g from pasteurised milk associated with 280 food poisoning cases in the Netherlands in 1989.

Mycobacterium Avium Subsp. Paratuberculosis

MAP is the causative organism of Johne's disease in cattle, a chronic wasting disease, and may occasionally be present in raw milk. Evidence linking MAP to a chronic inflammatory bowel condition in humans, called Crohn's disease, is becoming increasingly compelling. Concerns have been raised that MAP might be able to survive pasteurisation if present at levels above 100 cells per ml, especially if clumps of cells are present, and that pasteurised milk may therefore be a vehicle for Crohn's disease.

On the basis of new heat-resistance studies, many UK dairies have increased pasteurisation times to from 15 to 25 seconds. A survey of the level of contamination of pasteurised milk by MAP over a 17 month period, in 1999-2000, revealed a mean of 1.6% of raw and 1.8% of pasteurised samples were positive for MAP cultures indicating that commercially pasteurised milk may occasionally contain low levels of viable MAP. The potential public health impact of this situation is, however, still uncertain given that an association with Crohn's disease in humans remains unproven.

Viruses

A number of viruses have been shown to be present in raw milk, although many of these, such as Foot and Mouth Disease Virus (FMDV), are not pathogenic to humans. However, raw milk has been implicated in outbreaks of hepatitis and poliomyelitis. Some viruses, including poliovirus, are completely inactivated by pasteurisation, but this seems not to be the case with others, such as FMDV, if the virus is naturally present rather than inoculated. There is therefore the possibility that other viruses pathogenic to humans may survive at low levels, but, in bulk milk processing systems, it is thought unlikely that sufficient viruses will be present to infect consumers.

Toxins

Mycotoxins may be present in milk as a result of the ingestion of mouldy and contaminated feed by cattle. Feed contaminated by aflatoxin B1 as a result of the growth of *Aspergillus flavus* or *Aspergillus parasiticus* has been shown to give rise to the presence of aflatoxin M1 in the milk of dairy cows consuming it. However, only a small percentage of the ingested toxin appeared in the milk. Aflatoxins are persistent compounds and are not greatly affected by milk processing, and could therefore be present in pasteurised, packaged milk. However, recent surveys suggest that contamination of the milk supply is very limited and well within acceptable levels.

COAGULATION OF THE MILK

The principle of cheese processing is based on the coagulation of the protein in milk, during which about 90% of the milk fat is encapsulated. The coagulated mass is called curd, the remaining liquid is called whey. Curd consists mainly of milk proteins (casein) and milk fat; while whey mainly contains water, milk sugar (lactose), protein (serum proteins) and B-vitamins.

There are two basic ways to cause milk to coagulate:

1. Using an acid
2. Using a rennet (enzyme coagulation)

Acid Coagulation

This is mainly used to make fresh cheese. The acid can come from bacteria from the starter culture or from an acid added to the milk. When using a culture, pasteurised milk is inoculated. Inoculating agents can be a specific cheese starter culture, whey or sour milk. Curdling time depends on the amount of inoculating agent added (0.1-5% of the milk used to make cheese), temperature (20-35°C) and the starter culture used. It takes 2-16 hours. When the curd is firm, the curdling is finished. When using an added acid, you can use pure vinegar acid, lactic acid, citric acid or any other harmless organic acid. Sometimes a natural acid such as lemon juice is added. The acid can be added drop by drop to warm milk (about 80-90°C, just after boiling). Milk curdles much quicker at a high temperature. The curd is collected by straining the curdled milk through a coarse cloth. Sometimes the collected curd is pressed a little bit, and sometimes no pressure is applied. These products are consumed fresh.

Rennet

Rennet used in cheese processing can be of animal, vegetable or microbial origin.

The rennet in cheese has two functions:

- Rennet causes the milk to coagulate.
- During ripening, rennet causes milk protein to break down, giving the typical cheese taste.

Rennet is available in liquid form and as a dried powder. As it is an enzyme and therefore a biological product, the strength of the liquid product decreases during storage. Therefore, it is preferable to use the dried form. The concentration of the rennet is written on the package.

The amount of rennet to be added depends on the strength of the rennet and on the kind of cheese you are going to make. When making soft cheese, sometimes no rennet is needed; if rennet is added, only a little is needed (0.1 ml of rennet to 10 litres of milk). For hard or semi-hard cheeses, about 1.5 ml of rennet is added to 10 litres of milk; assuming the rennet concentration is 1:10,000. A starter is almost always added before or while adding rennet.

A number of factors influence coagulation:

- Amount of rennet or acid added.
- *Curdling Temperature*: When using rennet a small rise in temperature (for example from 30°C to 33°C) can significantly reduce coagulation time.
- Intensity of pasteurisation. Curdling is reduced when milk has been strongly heated. This effect can be neutralised by adding a small amount of CaCl2; for example 7 grams of CaCl2 per 100 litres of milk.
- *Fat Percentage of the Milk*: A high fat percentage means more fat must be encapsulated and curdling occurs a little bit slower.

Preparation of Rennet

Rennet from animal origin is prepared from the stomach of suckling calves or lambs. It is obtained by extraction of the stomachs in a salt solution with a preservative.

A traditional method formerly used in The Netherlands can be applied in the tropics:

- Take 9 litres of boiled water.
- Add 26 carved calf stomachs, 500 gram sodium chloride (salt) and 200 gram boric acid.
- Store these for at least 10 days and stir once per day.
- After ten days, remove the stomachs from the solution, add another 500 gram salt to the solution and store it for some days more.
- Then, pour it into clean bottles, close these and store the bottles in a cool and dark place. Your rennet is ready.

You have to try out the strength of the rennet. Also, the keeping quality of these rennets is usually poor. Sometimes the stomachs of the young animals are kept after cleaning, salting and drying. The dried stomachs are easy to transport and are used by some nomadic tribes. In cheese making a piece of dried stomach is then added to the milk.

SEPARATING CURD AND WHEY

Three methods can be followed to separate curd from whey:

1. Hanging the curd/whey mixture up in a clean cloth;
2. Putting the curd/whey mixture in cheese moulds, or in cylindrical forms with perforated sides;
3. First cutting and stirring the curd/whey mixture, putting the curd in the cheese moulds and then pressing the cheese.

When making fresh cheese, the first or second method is usually used. As the whey drains through the cloth or the mould, the volume of the curd will reduce to 1/2 or 1/3. When making ripened cheese, the last method has to be used to remove sufficient whey from the curd.

THE USE OF CHEESE WHEY

Whey is the by-product of cheese making. The acidity and the composition vary widely with the type of cheese and manufacturing process used. Whey from rennet-coagulated cheeses is less acid than whey of acid-coagulated cheeses. Also, the content of solids varies depending on the method of cheese making. If the content of solids is high, *e.g.* in the case of the processing of sheep cheese or in the case of rough curd treatment, the solids of the whey can be processed to a whey cheese. A rather well-known type of whey cheese is the Italian product Ricotta.

Manufacture of Ricotta from Cheese Whey

- Heat the acidified whey till at least 85°C or boil it for some time till the whey proteins have coagulated.

- Add some salt (0.1%) if desired.
- Collect the coagulated curd by filtering the mass through a towel or a filter.
- If you press the curd you will get a hard cheese; if the curd is pressed lightly or not at all, the cheese is consumed fresh.
- Add some salt if you like (if you did not do so before)
- Store the cheese in the fridge.

Whey as Animal Feed

If whey is discarded it can cause serious problems as a dairy effluent. Therefore it is far better to use the whey as an animal feed. It has a good feeding value because of the presence of some of the whey proteins. It can be used for feeding pigs or young animals like calves or lambs. Even adult cows or fattening cows can be fed with whey. It is important that the whey used for animal feed is acid (completely acidified) at the moment of feeding, otherwise the animals can get intestinal problems because of the lactose content of the whey.

COLLECTION AND PRESERVATION OF THE CURD

The amount of whey (water) in the curd has a great influence on the cheese properties. Maturation time, flavour, consistency, keeping quality of cheese, etc., all depend on the water content. Whey contains milk sugar or lactose. This sugar has to be converted into lactic acid by the lactic acid bacteria of the added starter. If a great deal of the whey is removed from the curd, you get cheese with little moisture, therefore a dry and hard cheese, which has to mature a rather long time. If little whey is removed from the curd, you get cheese with a lot of moisture, therefore a soft cheese. This cheese has in most cases a sour taste.

Since the curd treatment of fresh cheese and ripened cheese is quite different, they will be discussed separately. It is important during this step of chease making that the curd does not cool down and remains at about 30–36°C. The curd treatment can be started when the coagulated milk forms a firm mass. You can determine this by moving a small strip through the curd: if a smooth clear cut develops when the curd breaks it has coagulated sufficiently.

Curd Treatment for Fresh Cheese

As a rule, to make fresh cheese little whey is removed. After curdling the milk, the curd/whey mass is hung in a cloth or placed in cheese moulds. After 24 hours, enough whey has leaked out and the curd is cooled down. The cheese is then ready for consumption. The separation of the whey can be stimulated by piling the cloths or cotton sacks on top of each other so that more whey is pressed out of the cheese.

Curd Treatment for Ripened Cheese

After coagulation of the milk, the coagulated mass is cut with a sharp knife into square cubes of about 1.5 cm. It is important that the cubes are more or less of the same size and that the cutting is done very gently.

After cutting, the curd/whey mixture is briefly left to stand, for about 10 minutes, after which it is carefully stirred. After some time it can be stirred a little bit more vigourously. Most of the whey is removed from the cheese vat and the curd is placed in the cheese moulds. The curd is then pressed (0.1 kg/cm2). After about one hour the pressure can be increased to 0.4 kg/cm2. Other ways of determining the pressure are respectively 2 times and then 5 times the curd's own weight. After being pressed, the cheese is left for some time (4-20 hours) at 20-25°C.

In the meantime the lactose present in the cheese is completely converted into lactic acid. Then the cheese can be salted. Trying to produce cheese of a Gouda or Edam type in this way will be disappointing, because the cheese will become very firm, crumbly and acid. The reason for this is that the curd–after fermentation by the lactic acid bacteria-contains too much lactic acid.

Therefore Gouda cheese makers add 15 to 20% of hot water to the curd and whey after stirring for about 30 minutes. The temperature of the water is such that after addition of the water the temperature of the curd and whey is 35–36°C. (Water temperature is about 65–70°C. Before adding the hot water, part of the whey (30%) is drained off). After adding hot water the curd-and-whey mixture has to be stirred again for at least 20 minutes.

Salting the Cheese

Salting can be done in several ways:

- Stir the salt through the curd; add 30 grams of salt per kilogram of curd. This reduces the effect of the bacteria of the starter, but potentially damaging micro-organisms are also inhibited at an early stage.
- After pressing the cheese, rub it in with salt. Rub both sides in and then turn the cheese. This must be done every day for 3 days; in total 20 grams of salt is needed per kilogram of cheese.
- Place the cheese in brine of 200 grams of salt per litre of water at 12-16°C. During this pickling, the cheese absorbs salt and secretes lactic acid. The brine becomes more sour and improves in quality; it should therefore not be thrown away. Salt must be added regularly to the brine; 20 grams of salt is removed from the brine by each kilogram of cheese. For a small cheese, *e.g.* less than 1 kg, pickling time is about 12 hours. During pickling, the cheese must be turned once.

BUTTER

You will need:

- Pasteurised cream, sour cream or sour milk
- A heat source
- A pan
- A thermometer
- Cold water
- Sour milk or a starter culture
- A container for churning
- A sieve
- A bowl
- A tray for kneading
- Wooden spoons
- Fine salt if available (optional)
- Packing material, *e.g.* greaseproof paper or a jar
- Clean water

Butter is made by churning one of the following products:

- Cream,
- Sour cream, or
- Sour milk.

If you do not have enough milk from which to skim off the cream, the milk can be soured and churned as a whole (provided that the fat content of the milk is rather high; above 4%). But sour cream is better to churn than sour milk. It is only possible to manufacture sweet cream butter if the production and handling of the milk, cream and butter are exceptionally hygienic and the cream has undergone after pasteurisation a cold treatment by storing it for at least 12 hours at temperatures below 10°C.

Heating and Souring

Heat the milk or cream to a temperature of 85°C. Let it cool down to 18°C as quickly as possible (use a thermometer) using cold running water on the outside of the pan. Add 10-30 ml (about 1-2 tablespoons) of fresh fermented sour milk or a starter culture to one litre of milk or cream and stir. After about 24 hours at 16 to 18°C, the mixture will become thick and sour enough to be churned.

Churning

During churning the cream, sour cream or sour milk will be mixed intensively with air. This process causes fat globules to flocculate (or stick together), producing butter and buttermilk. The simplest way to make butter from small quantities of milk is by using a bottle or a jar that can be covered with a well-closing lid or a simple bowl with beaters. If large quantities of milk or cream are available, you should consider acquiring a real churn. There are several types available. The churn tub is a simple method, which is often used in the tropics. A cheap and practical domestic churn is a glass pot with a paddle attached to a screw top. The paddle can be turned manually. This churn is difficult to clean. It is best to rinse it with water before use, in order to prevent the butter from sticking to the sides.

Churns should not be filled more than one third with soured milk or sour cream. Churn with a regular up and down or sideways movement. Stop churning when the butter particles reach the size of rice grains or peas and the buttermilk looks rather liquid. If, after 30 minutes, no grains have yet been formed, you can change the temperature by adding a little clean cold or warm water. The amount of added water should never be more than 25% of the total amount of churned cream or milk.

The butter particles will float to the top of the buttermilk, because butter is lighter than buttermilk. This makes it easier to separate the two products by pouring off the buttermilk through a coarse sieve. Never add too much water, otherwise the buttermilk will become too watery.

Washing

Washing of the butter grains is not necessary. However, if very clean water is available, it can improve butter quality.

Washing the butter can be done in two ways:

1. Fill the churn 2/5 full with clean, cold water. Wash the butter by churning it for about 3 minutes. This can be repeated, if necessary. The idea is to remove part of the remaining buttermilk from the butter particles. It is important to remove as much as possible in order to produce butter that can be kept for a longer period. After washing, the butter particles can be skimmed off or the buttermilk can be drained.
2. If small quantities are available, the butter particles can be washed using a sieve. Put the sieve on top of a bowl and pour the mixture through it. Make sure that during churning the butter does not become one big lump, otherwise it will be difficult to wash-in case you want to do this.

Salting (If Preferred)

Salting is not necessary for preservation; many people, however, like the taste of salted butter. The butter can be slightly salted by kneading in about 10 grams of salt to each kilogram of butter. Mix the butter again the next day in order to allow the salt grains to dissolve.

Kneading

Kneading the butter is important in order to get a nice, smooth product. It helps to distribute the moisture and this improves quality and shelf life, provided that kneading is done in a hygienic way. Use a clean, well-rinsed kneading board. Knead the butter with the back of two wooden spoons until drops of water and buttermilk are not perceivable any more and the butter has a nice, smooth surface. Remove drops of buttermilk during this process. Instead of a wooden spoon, a wet roller or bottle can be used. If none of these are available, just use clean hands to knead.

Storage

Butter should be stored in a cool, dark place. Put it in a pot or wrap it up in greaseproof paper or aluminium foil. After some time, one to two weeks, the surface of the butter can be covered with moulds. This mould formation can be partially prevented by sprinkling salt on the surface or by wrapping the butter air tight. Moulds only grow if oxygen is available. Butter can also be frozen. However, after defrosting the butter will rather soon have an off-taste. It is therefore a good idea to divide the butter into smaller portions before freezing. Salted butter is less suitable for freezing.

Remarks

- If milk has gone sour without the use of a starter, but it still tastes and smells fresh, it can still be churned.
- Churning can take from 5 to as much as 60 minutes. A number of factors may influence the time needed, for instance:
- Type of animal the milk came from
- The fat content of the cream
- Treatment of the cream; the cream must have had enough time (at least 12 hours at 10–18°C) before churning for crystallisation of part of the fat to take place
- Feed eaten by the animal; feed influences the crystallisation (= melting point) of the fat
- The temperature of the cream during churning. This last aspect is also dependent on the melting point of the fat. If the cream is too cold, the fat globules do not easily stick together and churning will take longer. If the temperature of the cream is too high, churning will go quickly and the butter grains will not stick together and butter will not be formed. A good churning temperature is 15–20°C.
- Butter has a limited shelf life. It can become mouldy or rancid. An unpleasant cheese-like flavour may develop due to the deterioration of its protein. An alternative is ghee, which does not spoil as quickly as butter. Ghee is produced by removing the last water remnants from butter by heating the butter and letting the water evaporate, or by melting butter and draining the water, which separates from the fat.

FERMENTED MILKS

DEFINITIONS

Fermented milks have been produced by traditional methods for many centuries, and there are several hundred such products recorded around the world. They are produced as a result of microbial 'souring' of milk, usually cows' milk, but also the milk of other species, including sheep, goat, horse and buffalo. Most are very similar, both in terms of their characteristics, and in the technology used to produce them. Many fermented milk products are distinguished only by their region of origin, and very few have become commercially important.

Interest in these products, particularly yoghurt, has grown rapidly since the development of flavoured and fruit yoghurts in Europe in the late 1950s, and more recently as a result of the growing demand for, and marketing of, fermented milks as healthpromoting foods. Fermented milks can be conveniently classified on the basis of the type of fermentation they undergo, as lactic, yeast-lactic and mould-lactic. Lactic fermentation products can be further classified, depending on the characteristics of the lactic microflora, as mesophilic, thermophilic and probiotic or therapeutic.

LACTIC ACID FERMENTATON

Lactic acid fermentation is caused by some fungi and bacteria. The most important lactic acid producing bacteria is Lactobacillus. Other bacteria which produce lactic acid include:

- leuconostoc mesenteroides
- pediococcus cerevisiae
- streptococcus lactis
- bifidobacterium bifidus.

Lactic acid fermentation is used throughout the world to produce speciality foods:

- Western world: yogurt, sourdough breads, sauerkraut, cucumber pickles and olives
- Middle East: pickled vegetables
- Korea: kimchi (fermented mixture of Chinese cabbage, radishes, red pepper, garlic and ginger)
- Russia: kefir
- Egypt: laban rayab and laban zeer (fermented milks), kishk (fermented cereal and milk mixture)
- Nigeria: gari (fermented cassava)
- South Africa: magou (fermented maize porridge)
- Thailand: nham (fermented fresh pork)
- Philippines: balao balao (fermented rice and shrimp mixture)

The presence of lactic acid, produced during the lactic acid fermentation is responsible for the sour taste and for the improved microbiological stability and safety of the food. This lactic acid fermentation is responsible for the sour taste of dairy products such as cheese, yoghurt and kefir.

Lactic acid fermentation also gives the sour taste to fermented vegetables such as traditionally cultured sauerkraut and pickles. The sugars in the cabbage are converted into lactic acid and serve as a preservative.

Yogurt fermentation

Yogurt is made by fermenting milk with friendly bacteria, mainly Lactobacillus bulgaricus and Streptococcus thermophilus. Yogurt fermentation was invented probably by accident by Balkan tribes thousands of years ago. Yogurt remained mainly a food of eastern Europe until the 1900s, when the biologist Mechnikov created the theory that lactobacillus bacteria in yogurt are responsible for the unusually long lifespans of the Bulgar people. The milk sugar or lactose is fermented by these bacteria to lactic acid which causes the characteristic curd to form. The acid also restricts the growth of food poisoning bacteria.

During the yogurt fermentation some flavours are produced, which give yogurt its characteristic flavour. Yoghurt can easily be made at home using a live yogurt as the starter culture. To make you own yogurt use the following process. Bring the milk (or soymilk) to boiling point and cool down to 40- 45°C. Pour this milk in a sterile container and and per liter milk about 100 ml live yogurt. Mix with a sterile spoon and incubate at 40-44°C during 4 to 6 hours or until the yogurt is set. Put the yogurt in the refrigerator. If you worked under hygienic conditions, you can use your own yogurt as a starter for your next batch.

Magou fermentation

Magou is very popular in South Africa, especially among the Bantu people. Magou is a lactic acid fermented porridge made from maize. To make magua a 10 percent maize meal slurry is cooked, cooled and inoculated with wheat flour, which contains the bacteria. Magou is also produced on industrial scale and is then packed in cartons. In the industrial process the magou is inoculated with lactobacillus delbreuckii cultures.

Kefir fermentation

Kefir fermentation is similar to yogurt fermentation. Yogurt is only fermented by bacteria but kefir fermentation involves the help of bacteria as well as yeasts. These yeast produce some alcohol and carbon dioxide, which gives kefir its typical fizzy aspect. Kefir is inoculated with special kefir grains. These grains are mixtures of bacteria and yeasts in a matrix of proteins, lipids and carbohydrates. Kefir fermentation is done at room temperature, which makes the process easier. On the other hand, not everyone likes the taste of kefir.

MESOPHILIC

The genera of microorganisms that fall into this category include *Lactococcus*, *Leuconostoc* and *Pediococcus*. The optimal growth temperature is between 25-30 °C.

Traditional or Natural Buttermilk

Traditional or natural buttermilk is made from the liquid produced during butter production using a starter culture mixture of *Lactococcus* spp. and *Leuconostoc mesenteroides* subsp. *cremoris*.

Cultured Buttermilk

Cultured buttermilk is also produced mostly using a mixed culture of *L. lactis* subsp. *lactis*, *L. lactis* subsp. *cremoris* and the flavour-producing organisms *L. lactis* biovar *diacetylactis* and *L. mesenteroides* subsp. *cremoris*. It is traditionally made from skimmed milk. Ymer is similar to cultured buttermilk, but differs in the sequence of the manufacturing stages.

Nordic Sour Milks

Nordic sour milks such as Filmjolk and Nordic ropy milk are made using slimeproducing *Bacterium lacticus longi*, a synonym of *Lactococcus* spp. The slimy or ropy consistency of the products is also attributed to Butterwort leaves, which are rubbed on the interior of the pails.

Cultured Cream

Cultured cream or sour cream is made using the same starter cultures as cultured buttermilk, but has a much higher fat content.

Miscellaneous Products

Miscellaneous products include a range of traditional products that depend on spontaneous fermentation by naturally present lactic acid bacteria in milk. Maziwa lala is made using the same starter culture mixture as buttermilk, but is then sweetened. Susa, made from camel's milk is fermented using hetero-fermentative mesophilic starter cultures. Lben is similar to buttermilk but its production involves spontaneous fermentation. The microflora of this product mainly consists of *L. lactis* biovar *diacetylactis*, *Leuconostoc lactis*, *L. mesenteroides* subsp. *cremoris* and *Leuconostoc mesenteroides* subsp. *dextranicum*; lactobacilli, yeast, mould and coliforms are also present.

THERMOPHILIC

This category encompasses those starter cultures whose growth optimum is between 37 and 45°C. The genera of microorganisms that fall into this category include *Streptococcus* and *Lactobacillus*.

Yoghurt

Yoghurt is a term used to describe a wide range of related products, which may be classified according to legal standards, gel type and whether or not they are flavoured or if they are subjected to a further process. The usual starter culture employed to produce yoghurt is a mixture of *Streptococcus thermophilus* and *Lactobacillus delbrueckii* subsp. *bulgaricus*.

Acid Buttermilk

Acid buttermilk, also known as Bulgarian buttermilk is made using *L. delbrueckii* subsp. *bulgaricus* as the starter culture. *Str. thermophilus* or a cream culture may also be included in the starter culture.

PROBIOTIC OR THERAPEUTIC

LAB such as enterococci, lactococci, propionibacteria, *Leuconostoc*, and pediococci are used as probiotics, but the principal organisms are of the bacterial genera *Lactobacillus* and *Bifidobacterium*.

Yakult

Yakult is a term for a group of therapeutic products originating from Japan. The starter culture used is *Lactobacillus casei* subsp. *casei*, an organism naturally present in the normal intestinal microflora of humans. The organism is a probiotic strain that is thought to have a beneficial effect on the host, by improving the intestinal microbial balance. The positive health benefits of probiotics are reported to be of particular value in the treatment of diseases that result in a disturbance of the intestinal microflora.

Acidophilus Milk

Acidophilus milk is a traditional therapeutic milk product popular in eastern Europe, but now attracting more attention elsewhere for its perceived beneficial properties. It may be made from skimmed or whole milk, and the starter organism is *Lactobacillus acidophilus.*

'Bio' Yoghurts

'Bio' yoghurts are made by very much the same process as traditional yoghurt, and are very similar products, but usually use a mixed starter culture consisting of probiotic strains. *Bifidobacterium* spp. are often used, especially *Bifidobacterium bifidum*, and *Bifidobacterium longum*, together with lactobacilli, such as *L. casei* and *L. acidophilus*. These organisms are all found in the normal intestinal microflora and are considered to have a beneficial effect on human health.

YEAST-LACTIC FERMENTATIONS

Mesophilic LAB, thermophilic LAB and yeast are the main fermentation genera.

KEFIR

Kefir is a rather foamy and effervescent fermented milk that contains about 1% lactic acid and 0.5-1.0% alcohol, and is popular in eastern Europe and Mongolia. The starter culture consists of small, white 'kefir grains', about 2-10 mm in diameter.

These grains contain a complex and quite variable microbial community, but little is known about how they develop. The grains usually contain LAB such as *Lactobacillus* spp. plus *Lactococcus* spp., *Leuconostoc* spp., and *Str. thermophilus*, acetic acid bacteria, contaminants such as mould, and a number of yeast species such as *Saccharomyces and Kluyveromyces*, but the principal yeast species present is *Candida kefir*.

KOUMISS

Koumiss is traditionally made in central Asia from mares' milk, but is now often made from skimmed, or whole cows' milk with added sugar. Starter cultures contain LAB such as lactobacilli, strains of lactose-fermenting yeasts, non-lactose-fermenting and non-carbohydrate-fermenting yeasts. The finished product contains lactic acid, alcohol and carbon dioxide, producing a slightly effervescent drink.

MISCELLANEOUS PRODUCTS

Miscellaneous products such as acidophilus-yeast milk fall under the yeast-lactic group of fermented products, but little is known about the technology of these beverages.

MOULD-LACTIC FERMENTATIONS

Mesophilic LAB and mould are the genera responsible for fermentation.

VILLI

Villi is a fermented milk product from Finland, which is made from whole milk, using a starter culture of *L. lactis* subsp. lactis biovar *diacetylactis*, *L. mesenteroides* subsp. *cremoris*, and the mould *Geotrichum candidum*. The mould grows on the layer of fat that forms on the top of the product and produces a felt of mycelium.

INITIAL MICROFLORA

The initial microflora of fermented milk products is determined largely by the microflora of the whole and skimmed milks from which they are made.

PROCESSING AND ITS EFFECTS ON THE MICROFLORA

Although there is a very wide range of fermented milk products, the manufacturing technology used is generally very similar. The principal differences are in the starter cultures used, the composition and treatment of the milk, and the fermentation conditions. Therefore, for the purposes of this chapter, yoghurt manufacture is used as a representative example of fermented milk processes, since yoghurt is the most commercially important of these products. An outline of the process is depicted in Figure.

Several different varieties of yoghurt are produced, but the three main types are: set; stirred; and drinking yoghurt. Yoghurt is most commonly made from cows' milk, but may also be produced from the milk of sheep, goats, and, occasionally, other animal species. The composition of yoghurt varies slightly, and in some countries is regulated by legislation. Both whole milk and reduced-fat milks are used to produce yoghurt, but reduced-fat products have the largest market share in most countries.

A fat content of approximately 1.5% is typical for a low-fat yoghurt, and the milk is usually standardised to control the final fat content. The protein composition and quality of the milk are important, since they may have a significant effect on texture. Only milk of good microbiological quality should be used, in order to avoid problems of proteolysis associated with bacterial activity, and the production of bacterial proteases by psychrotrophs.

These enzymes may significantly alter the physical properties of the yoghurt, and cause defects. The milk-solids-not-fat content of the milk is usually increased to give a higher viscosity in the finished product. This may be done by fortification with non-fat dried milk or other dairy powders, or by concentration methods, such as evaporation under vacuum, or by membrane filtration. For most yoghurt, an MSNF level of about 15% is typical, but for drinking yoghurt, levels of less than 11% are preferred.

The milk is usually then filtered, to remove undissolved particles, de-aerated to provide conditions that favour rapid starter growth, and homogenised to improve texture and help prevent syneresis. Stabilisers may also be added to stirred yoghurts to improve viscosity and further reduce the likelihood of syneresis. Pre-gelatinised starch or plant gums are the most commonly used stabilisers.

HEAT TREATMENT

A heat treatment is generally applied to milk for yoghurt manufacture. A process of 80-85°C for 30 minutes is typical for batch processes, but, for continuous processes, a heat treatment of 90-95°C for 5-10 minutes is more usual. In some cases, a full ultra high temperature process may be applied. This relatively severe heat treatment has a number of effects.

Vegetative bacterial cells, which may include pathogens such as *Salmonella*, are killed, leaving only heat-resistant bacterial spores. Non-pathogenic organisms that might interfere with the growth of the starter culture are therefore reduced to very low levels. The heat treatment also has a significant effect on the final viscosity of the yoghurt, and improves texture by causing denaturation of the whey proteins and the three-dimensional aggregation of casein molecules.

The oxygen concentration of the milk is also further reduced, improving conditions for starter culture growth, since the starter organisms are generally microaerophilic. Finally, starter culture activity can either be stimulated or inhibited due to the breakdown products of heat-damaged milk proteins.

FERMENTATION

After heat treatment, the milk is cooled to 30 or 45°C and is then inoculated with the starter culture. Most commercial yoghurt production now uses a mixed inoculum containing defined strains of *Str. thermophilus* and *L. delbrueckii* subsp. *bulgaricus*. During fermentation, the two starter organisms grow synergistically. Initially, *Str. thermophilus* grows rapidly and produces lactic acid. As the acidity rises, so *L. delbreuckii* subsp. *bulgaricus* becomes more active and produces more acid along with aroma compounds.

It has been found that the growth of *Str. thermophilus* stimulates the growth of the *Lactobacillus*, probably by the production of a growth factor, thought to be formic acid, which stimulates proteolytic activity. *Lactobacillus* growth is further stimulated by the production of low levels of carbon dioxide from urea. The lactobacilli, in turn, stimulate the growth of *Str. thermophilus* by releasing peptides and, to lesser extent, amino acids as by-products of proteolysis.

The specific strains used are chosen for their effect on product flavour and texture, and strains may be used in rotation to prevent occasional problems with bacteriophage infection. Good hygienic practices after pasteurisation are required to prevent build-up of bacteriophage pools in stagnant whey. It is important that the inoculum contains a balanced population of the two organisms, usually a 1:1 ratio. Both liquid and freeze-dried cultures are used. Most commercial manufacturers use an inoculum of about 2-3% v/v.

Incubation of the inoculated milk at 40-45°C produces a rapid fermentation, which is normally complete within 3-4 hours. During this time, lactic acid is produced, giving an eventual acidity of 0.9-0.95% and a pH of approximately 4.6-5.0. The starter organisms may also produce aroma compounds, such as acetaldehyde, acetone, acetoin and diacetyl, and exopolysaccharides, which improve texture and viscosity, although if too much extracellular material is produced, a 'ropy' texture fault may result. Set yoghurt is fermented in the final container, but stirred yoghurt is fermented in bulk, and stirred slowly, for only a few minutes, during the process.

COOLING AND PACKING

When fermentation is complete, the yoghurt is initially cooled to about 15-20°C, to minimise further acid production. At this point, sweeteners, flavours and/or fruit purees may be added. These additions must be of good microbiological quality, since, for most yoghurts, no further processing is applied. The product is then dispensed into the final containers and further cooled to < 5°C.

A shelf life of about 3 weeks at this temperature is typical, although acid continues to be produced during this time and may affect flavour. Some yoghurt is heat treated after fermentation to destroy starter organisms, and this increases shelf life to several months. Greek-style, concentrated yoghurts are produced by further separation of the yoghurt after fermentation to increase the fat and solids content. Yoghurt may also be frozen, or dried for use as an ingredient.

6

Drink of Milk Product Items

There is but one real beverage and that is water. The other so-called beverages are foods, stimulants or sedatives. Milk is a rich food, one glass having as much food value as two eggs. Coffee, tea, chocolate and cocoa are stimulants, with sedative after-effects. Their food value depends largely on the amount of milk, cream and sugar put into them. Chocolate and cocoa are both drugs and foods. Alcohol is a stimulant at first, afterwards a sedative, and at all times an anaesthetic. When we think of drinking for the sake of supplying the bodily need of fluid, we should think of water and nothing else. If other liquids are taken, they should be taken as foods or drugs. Water is the best solvent known.

The alchemists of old spent much time and energy trying to find the universal solvent, believing that thereafter it would be easy to discover a method of making base metals noble. But they never found anything better than water. Water is the compound that in its various forms does most to change the earth upon which we live, and it is more necessary for the continuation of life than anything else except air. Pure water does not exist in nature, that is, we have never found a compound of the composition H_2O. Water always contains other matter.

The various salts are dissolved in it and it absorbs gases. The nearest we come to pure water is distilled. Pure water is an unsatisfied compound, and as soon as it is exposed it begins to absorb gases and take up salts and organic matter. Pure water differs from clean water. Clean or potable water is a compound which contains a moderate amount of salts, but very little of organic matter. Bacteria should be practically absent. Water that contains much of nitrogenous substances is unfit to use. If the water is very hard, heavily loaded with salts, it should not be used extensively as a drink, for if too much of earthy and mineral matter is taken into the system, the body is unable to get rid of all of them. The result is a tendency for deposits to form in the body. In places where the water is excessively charged with lime it has been noticed that the bones harden too early, which prevents full development of the body.

If the bones of the skull are involved, it means that there will not be room enough for the brain. Such diseases are rare in this country, but in parts of

Europe they are not uncommon. If the water is very hard, a good plan is to distill it and then add a little of the hard water to the distilled water. People who partake of an excessive amount of various salts can perhaps drink distilled water to advantage, but those who take but a normal amount of the salts in their foods should have natural water. Water forms three-fourths of the human body, more or less. It is needed in every process that goes on within the body. "To be dry is to die."

Water keeps the various vital fluids in solution so that they can perform their function. Without water there would be no sense of taste, no digestion, no absorption of food, no excretion of debris, and hence no life. The water is the vehicle through which the nutritive elements are distributed to the billions of cells of the body, and it is also the vehicle which carries the waste to the various excretory organs.

We can live several weeks without food, but only a few days without water. Hot water and ice-cold water are both irritants. Water may be taken either warm or cool. It is best to avoid the extremes. The amount of water needed each twenty-four hours varies according to circumstances. Two quarts is a favourite prescription. Those who eat freely of succulent fruits and vegetables do not need as much as those who live more on dry foods. Salt in excess calls for an abnormal amount of water, for salt is a diuretic, robbing the tissues of their fluids and consequently more water has to be taken to keep up the equilibrium.

Naturally, more water is required when the weather is hot than when it is cool. On hot days warm water is more satisfying and quenches thirst more quickly than ice water. Warm water also stimulates kidney action, which is often sluggish in summer. Ice water is the least satisfactory of all, for the more one drinks the more he wants. A normal body calls for what water it needs, and no more. An abnormal body is no guide for either the amount of food or drink necessary. Many people do not like the taste of water, especially in the morning.

This means that the body is diseased. To a normal person cool water is always agreeable when it is needed, and it is needed in the morning. People with natural taste do not care for ice water, but other water is relished. The common habit of drinking with meals is a mistake. Man is the only animal that does this, and he has to pay dearly for such errors. Taking a bite of food and washing it down with fluid lead to undermastication and overeating, and then the body suffers from autointoxication. A mouthful of food followed by a swallow of liquid forces the contents of the mouth into the stomach before the saliva has the opportunity to act. The best way is to drink one or two glasses of water in the morning before breakfast. Partake of the breakfast, and all other meals, without taking any liquid. Sometimes there is a desire for a drink immediately after the meal is finished. If so, take some water slowly. If it is taken slowly a little will satisfy. If it is gulped down it may be necessary to take one or two glasses of water before being satisfied. Those who have a tendency to drink too much during warm weather will find very slow drinking helpful in correcting it. If there is any digestive weakness, the liquid taken immediately after a meal should be warm and should not exceed a cupful. Those with robust digestion may take cool water.

Cold water chills the stomach. Digestion will not take place until the stomach has reached the temperature of about one hundred degrees Fahrenheit again, and if the stomach contents are chilled repeatedly the tendency is strong for the food to ferment pathologically, instead of being properly digested. For this reason it is not well to drink while there is anything left in the stomach to digest.

As stomach digestion generally takes two or three hours at least, it is well to wait this long before taking water after finishing a meal, and then drink all that is desired until within thirty minutes of taking the next meal. If the thirst should become very insistent before two or three hours have elapsed since eating, take warm water. Those who eat food simply prepared and moderately seasoned are not troubled much with excessive thirst. Two quarts of water daily should be sufficient for the adults under ordinary conditions. Here, as in eating, no exact amount will fit everybody. Make a habit of drinking at least a glass of water before breakfast, cleaning the teeth and rinsing the mouth before swallowing any, and then take what water the body asks for during the rest of the day. Taking too much water is not as injurious as overeating, but waterlogging the body has a weakening effect.

To drink with the meals is customary, not because it is necessary, but because we have a number of drinks which appeal to many people. Water is the drink par excellence. A food-beverage that is used by many is cambric tea, which is made of hot water, one-third or one-fourth of milk and a little sweetening. Children generally like this on account of the sweetness. It may be taken with any meal, when fluid is needed, but the amount should be limited to a cupful. It is not well to dilute the digestive juices too much.

The water taken in the morning helps to start the body to cleanse itself. Water drinking is a great aid in overcoming constipation. Constipated people generally overeat. Less food and more water will prove helpful in overcoming the condition. Unfortunately for the race, we have accustomed ourselves to partake of beverages containing injurious, poisonous substances. Inasmuch as this is the place to discuss the drugs contained in coffee and tea, we shall take the liberty of dwelling upon other habit-forming substances. They are all a part of the drug addictions of the race. For scientific discussion of these various substances we refer you to technical works. Coffee, tea and chocolate contain a poisonous alkaloid which is generally called caffeine. The theine in tea and the theobromine in cocoa are so similar to caffeine that chemists can not differentiate them. These drinks when first taken cause a gentle stimulation under which more work can be done than ordinarily, but this is followed by a reaction, and then the powers of body and mind wane so much that the average output of work is less than when the body is not stimulated.

The temporary apparently beneficial effect is more than offset by the reaction and therefore partaking of these beverages makes people inefficient. Coffee is very hard on the nerves, causing irritation, which is always followed by premature physical degeneration. Experiments of late indicate that children who use coffee do not come up to the physical and mental standard of those who abstain. The effect on the adults is not so marked because adults are more stable than children.

Those who are not used to coffee will be unable to sleep for several hours after partaking of a cup. Some people drink so much of it that they become accustomed to it. Coffee is not generally looked upon as one of the habit-forming drugs, but it is. However, of all the drugs which create a craving in the system for a repetition of the dose, coffee makes the lightest fetters.

It is surprising how often health-seekers inform the adviser that they "can not get along without coffee". If they would take a cup a few times a year, it would do no harm, but the daily use is harmful to all, even if they feel no bad effects and make it "very weak," which is a favourite statement of the women. Smoking, drinking beer and drinking coffee have a tendency to overcome constipation in those who are not accustomed to these things, but their action can not be depended upon for any length of time and the cure is worse than the disease. Tea drinking has much the same effect as coffee drinking, except that it is decidedly constipating. Perhaps this is because there is considerable of the astringent tannin in the tea leaves.

Chocolate is a valuable food. Those who eat of other aliments in moderation may partake of chocolate without harm, but if chocolate is used in addition to an excess of other food, the results are bad. The chocolate is so rich that it soon overburdens some of the organs of digestion, especially the liver. The Swiss consume much of this food and it is valuable in cases where it is necessary to carry concentrated rations.

Alcohol in some form seems to have been consumed by even very primitive people as far back as history goes. The Bible records an early case of intoxication from wine, and beer was brewed by the ancient Egyptians.

So much has been consumed that some people have a subconscious craving for it. There are cases on record where the very first drink caused an uncontrollable demand for the drug. Fortunately these cases are very rare. Alcohol is really not a stimulant, though it gives a feeling of glow, warmth and well-being at first, but this is followed by a great lowering of physical power, which gives rise to disagreeable sensations. Then the drinker needs more alcohol to stimulate him again. Then there is another depression with renewed demand: There is no end to the craving for the drug once it has mastered the individual. The lungs, heart, digestive organs, muscles, in fact, every structure in the body loses working capacity. Alcohol seems to have a special affinity for nervous tissue. A glass of beer or wine taken daily is no more harmful than a cup of coffee per day, but the coffee drinker does not make of himself such a public nuisance and menace as the man often does who drinks alcohol to excess. Formerly, it was respectable to drink. Some of our most noted public men were drunkards. Now a drunkard could not maintain himself in a prominent public position very long. To drink like a gentleman was no disgrace. Now real gentlemen do not get drunk.

In backward Russia they are becoming alarmed about the inroads of vodka, and are trying to decrease its consumption. France is trying to teach total abstinence to its young men because it disqualifies so many of them from military service to drink. Scandinavia is temperance territory. The German Kaiser has recently given a warning against drinking. The United States discourages drinking in the army and navy. Field armies are not supplied with alcoholics. Drinking is becoming disreputable.

It is very difficult to prove the harm done by excessive drinking of tea and coffee, also by the use of much tobacco, even if we do know that it is so. Everyone knows something about the deleterious effect of alcohol upon the consumer. Solomon wrote: "Wine is a mocker, strong drink is raging, and whosoever is deceived thereby is not wise. Who hath wounds without cause? Who hath redness of eyes?" Alcohol permanently impairs both body and mind. Depending on how much is taken, it may cause various ills, ranging from inflammation of the stomach to insanity. It reduces the power of the mind to concentrate and it diminishes the ability of the muscles to work. It reduces the resistance of the body and shortens life. Its first effect is to lull the higher faculties to sleep.

Most drunkards do not recover from their disease, for drunkenness is a disease. The various drugs given to cure the afflictions are delusions. Strengthening the body, mind and the will and instilling higher ideals are the best methods of cure. Suggestive therapeutics, and the awakening of a strong resolve for a better life are powerful aids. Proper feeding should not be overlooked, for bad habits do not flourish in a healthy body.

Civilization necessitates self-control and considerable self-denial. Those who go in the line of least resistance are on the road to destruction. It is often necessary to overcome habits which produce temporary gratification of the senses. Warden Tynan of the Colorado Penitentiary, 96 per cent of the prisoners are brought there because they use alcohol. It is also well known that moral lapses are most common when the will is weakened through the use of liquor.

Those who have the welfare of the race at heart are therefore compelled to give considerable thought to this subject. Past experience, it will not help to try to legislate sobriety into the people. Education and industrialism are the factors which it seems to me will be most potent in solving the alcohol problem. Morality, which in the last analysis is a form of selfishness, will teach many that it is poor policy to reduce one's efficiency and thereby reduce the earning capacity and enjoyment of life.

More and more the employers of labour will realise that the use of alcohol decreases the reliability and worth of the worker. Many will take steps like the following: "In formal recognition of the fact, established beyond dispute by the tests of the new psychology, that industrial efficiency decreases with indulgence in alcohol and is increased by abstinence from it, the managers of a manufacturing establishment in Chester, Penn., have attacked the temperance problem from a new angle. "Unlike many railways and some other corporations, they do not forbid their employees to drink, but they offer 10 per cent advance in wages to all who will take and keep—the teetotaler's pledge. Incidentally, a breaking of the promise will mean a permanent severance of relations, but there is no emphasizing of that point, it being confidently expected that the advantage of perfect sobriety will be as well realised on one side as on the other."

Business has during the past two centuries been the great civilizer, the great moral teacher. It has found that honesty and righteousness pay and that injustice is folly. Business has led the way to the acceptance of a new ethics, and new morals. What has been said about alcohol applies to tobacco in a much smaller degree. The use of tobacco seems to lead to the use of alcohol. It retards the development of children. It is surely one of the causes of various diseases. Tobacco heart, sore throat and indigestion are well known to physicians. Tobacco contains one of the deadliest of poisons known. One-sixteenth of a grain of nicotine may prove fatal. The reason there are so few deaths from acute tobacco poisoning is that but very little of the nicotine is absorbed. Men who chew tobacco make themselves disagreeable to others. Smoking of cigarettes is to be condemned not only because it poisons the body, but causes inattention and inability to concentrate on the part of the smoker, as well.

Every little while he feels the desire to take a smoke, and if smoking is forbidden he devises means of getting away. He robs his employer of time for which he is paid and injures himself. The ability to work is decreased by indulgence in smoking. Recent experiments show that for a short time there is increased activity after a smoke, but the following depression is greater than the stimulation, so there is an actual loss.

A few years ago, according to Mr. Wilson, who was then Secretary of Agriculture, there were about 4,000,000 drug addicts or "dope fiends" in the United States. Without doubt this estimate was too high, for the proportion of addicts in the country is not as great as in the large cities. The drugs chiefly used are cocaine, opium, laudanum, morphine and heroin. These drugs are much more destructive than alcohol. Cocaine and heroin are the worst. It is very difficult to stop using any of them once the habit has been formed. Nearly every "fiend" dies directly or indirectly from the effect of his particular drug. Every one weakens the body so that there is not much resistance to offer to acute diseases. Every one destroys the will power so that a cure is exceedingly difficult.

It is well to bear in mind that all are not possessed of strong enough will power to resist their cravings and that some take to cocaine when they can not get liquor. Cocaine is far worse than alcohol. People should be very careful about taking patent medicines. There is no excuse for taking them. The most popular ones have as their basis one of the habit-forming drugs. Most of the soothing syrups contain opium in some form. To give babies opiates is a grave error, to speak mildly. It weakens the child, may lay the foundation for a deadly habit later in life, and often an overdose kills outright. Well informed mothers avoid such drugs and keep their children reasonably quiet by means of proper care. Many of the remedies for nasal catarrh and hay fever contain much cocaine. Cocaine is an astringent and a painkiller and people mistake the temporary lessening of discharge from the nose and disappearance of pain for curative effects. But there is nothing curative about it. In a short time the mucous membrane relaxes again and then the discharge is re-established. The nerves which were put out of commission resume their function and then the pain reappears.

Opium or one of its derivatives is generally present in the patent medicines given for coughs. Opium is also an astringent and will suppress secretions, but this is not a cure. Excessive secretions are an indication that the body is surcharged with poison and food. Let them escape and then live so that there will be internal cleanliness and then there will be no more coughs and colds. The unfortunate people who get into the habit of using these drugs degenerate physically, mentally and morally. They need more and more of their drug to produce the desired effect until they at last take enough daily to kill several normal men. Sometimes they are able to keep everybody in ignorance of what they are doing for years. They develop slyness and secretiveness.

They become very suspicious. They are nearly always untruthful, and those who deal with them are surprised and wonder why those who used to be open and above-board now are furtive and dishonest. They often lie when there is not the slightest excuse for it. The moral disintegration is often the first sign noticed. After habitually using any of these drugs for a while the body demands the continuation and if the victim is deprived of his accustomed portion there will be a collapse with intense suffering. Every tortured nerve in the body seems to call out for the drug. The victim will do anything to get his drug. He will lie, steal, and he may even attack those who are caring for him. For the time being he is insane. Many professional men use cocaine. It is a favourite with writers. It often shows in their work. Those who write under the inspiration of this drug often do some good work, but they are unable to keep to their subject.

Their writings lack order. We have enough of such writings to have them classified as "cocaine literature". If there are 4,000,000, or even fewer, of these people in our land, it is a serious problem, for every one is a degenerate, to a certain degree. If the medical profession and the druggists would cooperate it would be easy enough to prevent the growth of a new crop of dope fiends. Of course, people would have to stop taking patent medicines, which often start the victims on the road to degeneration.

Then the physicians should stop prescribing habit-forming drugs, as well as all other drugs, and teach the people that physical, mental and moral salvation come through right living and right thinking. Unfortunately the medical profession is careless and is responsible for the existence of many of the drug addicts. A patient has a severe pain. What is the easiest way to satisfy him? To give a hypodermic injection of some opiate. The patient, not realizing the danger, demands a pain-killer every time he suffers. He soon learns what he is getting and then he goes to the drug store and outfits himself with a hypodermic outfit and drugs, and the first thing he knows he is a slave, in bondage for life.

This is no exaggeration. There are hundreds of thousands of victims to the drug habit who trace their downfall to the treatment received at the hands of reputable physicians, who do not look upon their practice with the horror it should inspire because it is so common. Doctors do not always bury their mistakes. Some of them walk about for years. In spite of laws against the sale of various drugs, they can be obtained. There are doctors and druggists of easy conscience who are very accommodating, for a price. There is no legitimate need for the use of one-hundredth of the amount of these drugs that is now consumed. A local injection of cocaine for a minor operation is justifiable, but none of the habit-forming drugs should be used in ordinary practice to kill pain, for the proper application of water in conjunction with right living will do it better and there are no evil after effects. Massage is often sufficient.

To show a little more clearly how some people become addicted to drugs, let us consider one of the latest, heroin: A few years ago this drug, which is an opium derivative, was practically unknown. It is much stronger than morphine and consequently the effect can be obtained more quickly by means of a smaller dose. Physicians thought at first that it was not a habit-forming drug, for they could use it over a longer period of time than they could employ morphine, without establishing the craving and the habit.

So they began to prescribe heroin instead of morphine, and many a morphine addict was advised to substitute heroin. All went well for a short while, until the victims found that they were enslaved by a drug that was even worse than morphine. Now, thanks chiefly to the medical profession, it is estimated that we have in our land several hundred thousand heroin addicts. Sallow of face, gaunt of figure, looking upon the world through pin-point pupils, with all of life's beauty, hope and joy gone, they are marching to premature death. The medical profession furnishes more than its proportion of drug addicts. They know the danger of the drugs, but familiarity breeds contempt. If the public but knew how many of their medical advisers, who should always be clear-minded, are befuddled by drugs, there would be a great awakening. One eminent physician who has now been in practice about forty-five years and has had much experience with drug addicts, has said that according to his observations, about one physician in four contracts the drug habit. I believe this is exaggerated, but I am acquainted with a number of physicians who are addicts.

Physicians who smoke do not condemn the practice. Those who drink are likely to prescribe beer and wine for their patients. Those who are addicted to drugs use them too liberally in their practice. Those who have watched the effects of the various drugs, from coffee to heroin, must condemn their use. It is true that an occasional cup of coffee or tea, a glass of wine or beer does no harm. A cigarette a week would not hurt a boy, nor would on occasional cigar harm a man.

But how many people are willing to indulge occasionally? The rule is that they indulge not only daily, but several times a day, and the results are bad. One bad habit leads to another, and the time always comes when it is a choice between disease and early death on one hand, and the giving up of the bad habits on the other, and when this time comes the bonds of habits are often so strong that the victim is unable to break them. We realise that knowledge will not always keep people out of temptation and that some individuals will take the broad way that leads to destruction in spite of anything that may be said. Youth is impatient of restraint and ever anxious for new experiences.

Regarding this serious matter of destructive drug use, much could be done by teaching people their place in society: That is, what they owe to themselves, their families and the public in general. In other words, teach the young people the higher selfishness, part of which consists of considerable self-control, self-denial and self-respect.

Drugs are too easy to obtain today. Some day people will be so enlightened that they will not allow themselves to be medicated. This is the trend of the times. Until such a time comes, society should protect itself by making it very difficult to get any of the habit-forming drugs. If necessary, the free hand of the physician should be stayed. Much of the confidence blindly given him is misplaced.

HEALTH RISKS OF CONSUMING DAIRY PRODUCTS

Most dairy products contain large amounts of saturated fat. Some dairy products may cause health issues for individuals who have a lactose intolerance and milk allergies. Some dairy products such as blue cheese may become contaminated with the fungus Aspergillus fumigatus during ripening, which can trigger asthma and other respiratory problems in susceptible individuals. Vegans and some vegetarians avoid dairy products due to a variety of ethical, dietary, environmental, political, and religious concerns.

MILK PRODUCTION

The bacteriological quality of raw milk at the time of milking in India is comparable with that in the advanced dairying nations. Subsequently, however, the quality deteriorates due to improper handling of milk and lack of availability of infrastructure like all-weather roads, cooling facilities, potable water, regular electric supply and sewage disposal. A holistic approach will be taken to address the issue of clean milk production, which is imperative for marketing and promoting export of dairy products. Steps will also be taken for development of unorganised milk sector that controls a significant portion of the liquid milk and sweetmeat market.

DEVELOPMENT OF A MARKETING NETWORK

The development of a marketing network and remunerative price support to the producers are great incentives for higher animal productivity and these will be encouraged for all types of livestock products. Even the advanced countries are giving direct and indirect price support to livestock farmers. Priority attention should also be given to improve processing, marketing and transport facilities for livestock products and value addition thereon.

External markets are an extremely important source of demand and these will be tapped much more aggressively. In order to encourage exports, licensing control for processing of livestock products/by-products will be repealed and restrictions on the export of livestock and its products will be removed.

The immediate focus will be on export of animal and poultry products to Asian and African countries. The minimum requirements for sustainable export are creation of disease-free zones, organic farming and potable water. These will be made available in selected areas having large marketable surplus. India has a large number of animal markets where livestock are traded but these are not developed on scientific lines. Market facilities are generally inadequate and, if available, are poorly maintained.

Development of organised markets with adequate facilities will, therefore, be taken up. The concept of organic farming can also be extended to animal products. Indian animals are reared in village pastureland and they are not generally treated with hormones, feed-antibiotics, or other drugs, so their products are healthy, wholesome and natural in every sense of the word. In rural India, cow dung and biomass are primarily used as manure. Initiative for export of 'Grassfed' animal products will be taken. Necessary infrastructure for certification procedures related to organic animal farming will be promoted.

BUTTERMILK AND SOUR MILK PRODUCTS

Buttermilk is a by-product of the butter-making process. The taste can be more or less sour depending on the sourness of the cream or the milk, which is used for butter making, or on the degree to which it sours after churning. It is also possible to make a product like soured buttermilk using milk or skimmed milk, by inoculating it with sour milk and letting it ferment for one day.

For the fermented milk you will need: fresh (skimmed) milk, a heat source, a wooden spoon, fresh fermented milk or buttermilk or a starter culture, a saucepan with a thick bottom, and a thermometer. Heat the fresh (skimmed) milk to boiling point, stirring all the time. Cool it down to 18–20°C, for instance in a large pan with cold water. Add 10–30 ml of sour milk or buttermilk or a starter culture per each litre of milk (1%). Leave for it 18-24 hours at room temperature (18–20°C); if the surrounding temperature is higher, fermentation time will be somewhat shorter. After this the sour milk is ready. Store it in a cold place (cool basement or a refrigerator), if you want to keep it for some days.

RABI

Rabi is sweetened, concentrated milk. During concentrating, sugar is added from time to time.

You will need:

- (Unboiled) milk
- A heat source
- A wide, shallow iron pan with a thick bottom
- A flat metal scoop
- Sugar
- Scales.

Add sugar to the milk during the heating process (maximum 300 g per litre of milk) and follow the same procedure as for koa. Lumps of sugar will often be found in the end product.

YOGHURT

Yoghurt is produced when milk is soured by certain lactic acid bacteria, which prefer growing temperatures far above room temperature: 37–45°C. The milk should first be heated to 85°C or higher. A high pasteurisation temperature (above 72°C) gives a better consistency (thickness) to the final product. After the milk has been soured, the resulting yoghurt can be used to make more fresh yoghurt by adding it to fresh milk.

Basic Recipe for Yoghurt

You will need:

- Fresh raw milk

- A heat source
- A saucepan
- A spoon
- A thermometer
- Cooling facility (*e.g.* a large pan with cold water)
- A cool place (refrigerator or cellar)
- Starter culture for yoghurt or some fresh yoghurt
- Thermos flask or a box covered with a blanket

Heat the milk to 85°C or higher and keep it at this temperature for 3 minutes. Cool the milk to 45°C. Add 30 ml (2-3 tablespoons) of fresh yoghurt to each litre of milk; the yoghurt should not be more than 2 days old. Instead of fresh yoghurt you can use a yoghurt starter culture. Mix the milk and the starter and leave it to ferment. The time required for the milk to turn sour depends on the temperature.

To give you an idea:

- At 40-45°C it takes about 3 to 6 hours
- At 35-37°C it takes about 20 to 15 hours
- At 30°C it takes about 24 hours

The ideal temperature to make pleasant-tasting yoghurt with a firm consistency is 40-45°C. It is not possible to produce yoghurt at temperatures below 30°C or above 50°C.

The correct temperature can be maintained using an insulated box or a blanket. Yoghurt is ready for consumption once the incubation period is finished. If cooled, yoghurt can be kept for one week.

Using a Thermos Flask

Heat the milk to at least 85°C, then cool it to 45°C. Pour 90% of the milk into a thermos flask, which has been rinsed with hot water. Mix 1-2 tablespoons of fresh prepared yoghurt (or yoghurt culture) with the rest of the milk and add this to the thermos flask. Close the flask well and leave to stand for 3-6 hours. Remove the yoghurt from the thermos flask and store in a cool place. Yoghurt made of sheep milk is very firm and therefore not suitable for fermentation in a thermos flask.

Yoghurt made from Milk Powder

Make milk from milk powder according to the instructions on the package, but add 10 to 15% extra milk powder. Dissolve the milk powder in water, heat it to boiling point and let it cool down to 45°C. Stir in 1-3 tablespoons of fresh yoghurt or a yoghurt culture per litre of milk. Cover the saucepan and put it in a warm, insulated place. After 3-6 hours this (firm and concentrated) yoghurt should be ready for consumption.

Remarks

- It is best to use fresh milk to make yoghurt. Milk powder can also be used. Sterilised milk may give a thinner yoghurt than pasteurised milk.
- After incubation, cooling is desirable, preferably below 10°C, so that souring is stopped (this retains the pleasant taste) and the bacteria remain more viable, allowing the yoghurt to be used to inoculate milk again.
- Make sure the milk ferments as quickly as possible, preferably at 40-45°C rather than 30°C. Harmful bacteria have less opportunity to develop if the fermentation process goes faster.
- Thicker yoghurt can be produced by adding 2 to 3 tablespoons of milk powder to each litre of milk before heating it to 85°C.
- It is not advisable to use fruit yoghurt from a shop as a starter culture because it contains a lot of additives. Plain yoghurt from a shop can be used if it is not too old. Sterilised yoghurt is not suitable either, because the yoghurt bacteria have been killed by the sterilisation process.
- When using yoghurt from a carton or a pot as a starter, first remove the top layer and take the yoghurt from the centre to make fresh yoghurt. This is because the bacteria in the middle are probably the most diverse and active.
- Stir the product as little as possible before removing some, to avoid the extra risk of incorporating undesirable bacteria.

GHEE

To make ghee, you will need:

- Butter
- A heat source
- A pan
- A metal spoon.

Heat the butter until water and fat form separate layers; the fat will float on top.

There are two ways to remove the water:

1. It can be removed by further heating. The water present will evaporate.
2. It is possible to remove the layer of fat with a spoon. This fat should then be heated again. The scum, which will form, has to be skimmed off regularly, preferably with a skimmer. The colour of ghee can vary from almost white to dark brown. A rancid flavour is acceptable, but if it tastes burnt it should be discarded.

KOA

Koa or khoa is a type of concentrated milk.

You will need:

- Fresh (unboiled) whole or skimmed milk
- A heat source
- A shallow, clean, wide, iron pan with a thick, flat bottom
- A flat, clean metal utensil to stir with (*e.g.* a flat pancake spatula).

Fill the pan with the milk up to 30 to 50% of its capacity. Bring the milk to boiling, stirring continuously. The water will evaporate and after some time the milk will reach a certain viscosity, this means the milk will thicken. Take great care to scrape the sides of the pan during stirring.

Once the milk has reached a dough-like consistency and stirring does not prevent the mass from sticking to the side of the pan, you can remove it from the heat source and lower the temperature by at least 20°C. At this point the water content should have been reduced to about 40% of what it was. Take the lump of koa out of the pan, put it on a cold surface and flatten it. After cooling, the koa will be firm and can be cut into squares. It will have a sweet, nutty flavour. Koa can be kept only for about 2-5 days because of possible re-infection. The preparation of koa requires much time (a few hours) and fuel, on top of which 1 litre of milk produces only 0.4 litres of koa.

KEFIR

Kefir, like yoghurt, is a milk product first discovered by the nomadic tribes living in the cold areas of the Caucasus. When making kefir, acid, gas and some alcohol are produced. Like sour milk and buttermilk, it has a special aroma which is different from that of yoghurt. Kefir is made by using a 'yoghurt plant', which is actually a misleading name as it is not a plant and it has nothing to do with yoghurt. The 'yoghurt plant' is in fact a cauliflower-like lump of chalky crystals and micro-organisms consisting of yeasts and bacteria.

The yeasts produce alcohol and gas, while the bacteria convert the milk sugar into lactic acid. You may be able to obtain a piece of this from someone who regularly makes kefir. If not, get some dried kefir granules at the local market. You will need: fresh raw milk, a saucepan, a heat source, a thermometer, a clean pan, a glass bottle with wide neck, a bottle which can be closed tightly, kefir granules or a 'yoghurt plant', a sieve, cooling facilities, clean water, a teaspoon, and a cool place to store the product.

Boil the milk and put it into a bottle, which has been thoroughly cleaned and rinsed with hot water. Do not fill the bottle completely; let the milk cool down to 20°C (use a thermometer). Add one tablespoon of kefir granules soaked in water to each half litre of milk in the bottle and cover it loosely so that the gas produced can escape. You can use a fresh yoghurt plant instead of the soaked granules. Keep the bottle at a temperature of 16 to 18°C. After

24 hours the milk will have become a little thicker, forming some froth: this is kefir. Sieve the kefir and use the (washed with clean water) granules, which remain in the sieve to make fresh kefir again. The kefir is now ready for consumption or can ripen for some days to get more flavour.

Ripening of the kefir

The kefir can be left to 'ripen', during which time further fermentation takes place. This is essential to create kefir's characteristic qualities. Pour the kefir into a well-cleaned bottle, which can be closed, or a bottle with a clip fastening and do not fill more than 3/4 because gas forms during ripening. Leave the bottle at about 15°C, but do not store it for more than 3 days.

Towards the end of the ripening process the whey separates and can be incorporated again by stirring or turning the bottle. The end product is a thick, creamy, frothy drink with a sour taste and smell and the produced carbon dioxide is perceivable. If the kefir is left to ripen for more than 3 days, the milk may curdle and the drink becomes too sour. The temperature and the time are important as they determine the flavour. The kefir can be kept in a refrigerator or a cellar for a few days.

Storage of the Kefir Granules

If kefir production is stopped for some time, the kefir granules can be dried as follows: Put them into a sieve and rinse well with clean water until all remaining milk is removed. Place them on a clean cloth and leave to dry in a clean place, but not in the sun, until the granules have shrivelled up. The granules can be kept in a closed bottle in a cool place for 12 to 18 months. The rinsed granules can also be kept in a freezer. Kefir granules can also be kept in a pot with water at 4°C, but they become inactive after 8 to 10 days.

Remarks

- There is a greater chance of failure with kefir than with yoghurt. The end product may not only taste bad, it can also be unhealthy. In order to produce good kefir, the appropriate hygiene should be applied. Care should also be taken not to work at too high temperatures.
- The shelf life of kefir is the same as for yoghurt, sour milk and buttermilk: its quality will rapidly decrease if it is stored too long. At 5°C, soured products can be kept for about ten days; at 10°C this is three days, and 20°C is too high.
- Kefir may be too sour or too yeasty (due to an incorrect balance between the activity of the bacteria and yeasts). A fishy or ammonia-like taste can be the result of a breakdown of proteins by undesirable bacteria, which enter due to insufficient hygiene.

CHEESE PRODUCTS

Cheese is a product that has been made for centuries. Virtually all the nutrients present in milk are concentrated in cheese. There is a huge assortment of cheeses. They can have various compositions; we make a rough distinction between fresh cheese and matured cheese, and between soft cheese and hard cheese. Fresh cheese can be consumed immediately after production, whereas matured cheese has to be stored after processing to develop flavour and a good consistency. Soft cheese has a higher water content than hard cheese; moreover hard (or semi-hard) cheese generally has a clean, dry rind. Soft cheeses and hard cheeses can be matured for some weeks or even up to a number of years.

Table. Rough Composition of Milk and Some Types of Cheese

Product	**Content (in% Weight) of**			
	Water	**Fat**	**Protein**	**Salt (Sodium Chloride, NaCl)**
Milk	87.5	3.5	3	less than 0.1
Skimmed milk	91	0.1	3	less than 0.1
Fresh cheese (whole fat)	73	10	10	less than 0.1
Fresh cheese (low fat)	83	0.2	13	less than 0.1
Cream cheese (fresh)	57	32	9	0.5
Soft cheese	51	25	19	2
Semi-hard cheese	42	30	25	2–2.5
Hard cheese	30	35	30	2–2.3

Cheese production in the tropics is typified by specific problems; consequently, production processes and end products differ from those in Western Europe:

- Outside the temperate zones, high temperatures and very high air humidity must be taken into consideration. Both these factors are unfavourable for cheese making, in particular for the maturation of cheese.
- Milk is often available in small quantities, while its quality may not be up to standard. In particular, its hygienic quality and composition may leave much to be desired.
- Hard and semi-hard cheeses require good-quality milk and hygienic processing. Moderate temperatures are needed for good ripening and storage of these cheeses. In warm and hot climates most cheeses are produced on a small scale and cannot be stored for a long time. Therefore generally these are soft cheeses.

- In many tropical and subtropical countries, very matured cheese is not very popular. Most consumers are not used to the distinct flavour and smell of such cheese.

The shelf life of cheese can vary from several days to some months and even to some years, depending on its processing. Fresh cheese can, like fermented products, be stored for only a short time; it must be consumed immediately or within a few days. The keeping quality of such cheese can be improved by cold storage and salting. Fresh cheese generally does not have a rind and is packed in a wrapping or a package.

Most of the fresh cheese has been curdled using only an acid; it cannot be ripened and it should be consumed within a couple of days. Soft cheese is processed by using acid and rennet, in most cases it matures for some weeks or even some months.

Soft cheese generally has a rind and often also a microbiological flora on the surface of the rind (like Camembert and Brie). Semi-hard and hard cheese can be kept well for 3-4 months or more. Cheese which is able to continue ripening should not be kept in a refrigerator, but in a cellar or cool place at 10-15°C.

Cheese making has three main goals:

1. Concentration of the milk into curd, whereby a great deal of the liquid part of the milk, the whey, is removed.
2. Preservation of the curd. Acidification of the curd and the salting process are important, along with a well-closed cheese rind or effective wrapping of the cheese.
3. Maturation of the cheese, in order to get a good taste, flavour and consistency.

The cheese-making process has some essential steps, these are:

- Coagulation of protein of the milk, whereby the fat is incorporated in the coagulated protein
- Draining of the whey
- Aacidification of the curd
- Collection of the curd and combined growth into a cheese
- Salting
- Maturation
- The processing of fresh cheese follows steps 1 to 3; matured cheese needs steps 1 to 6. The basic operations in cheese making will be discussed in this chapter: the treatment and quality of the milk used to make cheese (pasteurisation and/or standardisation)
- The coagulation (curdling) of the milk using acid or enzymes
- The separation of the curd and whey

- The collection and preservation of the curd (heating, collection, pressing and salting)
- The maturation (ripening) of the cheese.

UTENSILS FOR MAKING CHEESE

When making cheese, you will not always need all the utensils described below. Decide beforehand which you will use, and make sure that they are clean and rinsed with clean water. It is preferable to use utensils made of stainless steel or glass.

The utensils used during cheese making are:

- A thermometer with a range of 20–100°C
- A measuring cup
- A bucket for coagulation of the milk
- Cheese moulds, which can be made in various ways
- Cheesecloth, its size depends on the size of the mould
- Cutlery and other tools such as:
  - Spoons to measure rennet and/or acids
  - A knife to cut curdled milk
  - A skimmer to scoop curd out of whey, or a colander.

Cheese moulds can be made from various materials *e.g.* wood, plastic or metal. Metal materials should be non-rusting. Do not use plastic tubing used in construction as it can release poisonous substances. Cut the moulds lengthwise and drill holes from the inside out. Plastic moulds for the production of soft cheese are also commercially available. Wooden cheese moulds can be used as well. Cheese that will have a closed rind after pressing is never placed directly into a mould. Instead, a clean cloth of linen or cotton is first placed in the mould and it is then filled with curd. The cheesecloth is then folded closed.

To make hard or semi-hard cheese you will also need the following:

- A curd knife to cut the renneted milk
- A cheese press

Buy these second-hand if possible or make your own simple press, paying special attention to the following:

- The materials used must not be poisonous.
- It must be possible to easily clean the materials.
- It must be possible to build up sufficient pressure (0.1 to 0.4 kg/cm2 or 2-5 x the weight of the cheese).

THE QUALITY OF MILK USED TO MAKE CHEESE

Cheese making begins with hygienic milking. The quality of the milk used greatly influences the smell, taste and keeping qualities of cheese. The composition of milk can vary greatly. Just before a cow is set dry–the period in which a cow is not milked–and just after calving, the cow milk has a different taste and composition. Milk from a cow with an udder infection is not suitable for human consumption, and therefore cannot be used for making cheese.

Apart from the usual measures taken to ensure good hygiene during milking, attention must be paid to the following points:

- Clean the equipment very well.
- Disinfectants and cleaning agents must never be left in or on equipment. Disinfectants inhibit the growth of the bacteria of the starter or souring agent.
- The room in which cheese is prepared must be kept properly clean.
- The milk used for cheese making should preferably be pasteurised (15 seconds at 72°C or 30 minutes at 63°C). More intense heating is not desirable because it reduces the curdling of milk. More rennet or calcium chloride (CaC12) is then needed.
- For the above reasons, milk powder which is strongly heated during its production (high-heat milk powder) is less suitable for cheese processing. Only low-heat milk powder can be used.
- Sour milk neutralised with sodium bicarbonate will not curdle well.

MATURATION OF THE CHEESE

During the maturation of the cheese, protein and fat are partly broken down, causing a change in structure and flavour. The older the cheese, the more flavour it will have. Sometimes, after more than one year the structure becomes crumbly because of the protein breakdown. Maturation is caused primarily by the enzymes in the cheese. To mature cheese, it can be stored in a cool room.

This is a good method but technically rather difficult in the tropics. After salting or pickling, cheese is ripened on a wooden shelf in a cool environment (12-20°C) with a not too high relative humidity (about 80%). Cheese must be turned regularly: initially, *i.e.* during the first 2 to 3 weeks, daily and later once a week. If mould grows on the cheese, it can be removed with water or vinegar.

In addition to this method, the following treatments can be carried out on curd:

- Curd can be smoked over a fire.
- Curd can be boiled in brine.
- You can salt curd, make little balls of it and leave them to dry in thesun.

The products which result are particularly suitable to be processed in other foods. After boiling, cheese will not ripen anymore because the enzymes inside the cheese are destroyed by heating.

7

Components of Milk food

Milk contains components that are essential to humans such as proteins, carbohydrates, fat, water, all the B-vitamins, vitamins A and D, calcium and phosphorus. It also provides energy.

An important protein in milk is casein (in many cases 80% of the milk protein). This is the base for cheese making. Casein is linked to calcium phosphate, which is why milk contains a relatively large amount of this salt that is a very important nutrient for humans and animals. In addition to casein, milk contains whey proteins (20% of the milk protein). The whey proteins are in most cases not incorporated in the cheese; they remain in the whey. Whey proteins (globulins and albumins) have a very high nutritive value.

Milk protein is of a high quality. This means that the human body can use a large part of the protein efficiently. Proteins in various other foodstuffs have a comple-mentary effect. In combination with cereals, potatoes, meat, eggs or nuts in one meal, the body can use an even greater percentage of the milk protein. Apart from milk, there are other animal protein sources such as fish and meat.

Vegetable protein, which is also important in making the body's proteins, is found in cereals and pulses. Protein is needed by the body for growth, replacement of worn-out body proteins and the production of compounds that the body needs. Milk sugar (lactose) is a carbohydrate, a necessary component to keep the body going. Our bodies burn carbohydrates in the same way an oven burns wood. Through this combustion, energy is released which is used by our bodies for many kinds of activities.

Milk fat is present in the form of small fat globules, which have a lower weight than the other components of the milk. When cow milk is allowed to stand, these globules collect on top of the milk and form a layer of cream. Buffalo milk also forms some cream on top, but other kinds of milk, such as that of sheep and goats, hardly form a layer of fat at all. For these types of milk one needs to separate the cream from the milk. Milk fat is easy to digest. The body uses fat as a fuel or stores it as fat reserves. Milk is also an important source of minerals and vitamins. It contains large quantities of calcium, which can easily be absorbed by the body after digestion and is important for the formation of bones (the skeleton).

Milk is also an important source of vitamin B2 (Riboflavin), but there is little vitamin C in milk. Therefore a person's diet must also include vegetables and fruits in order to ensure a sufficient supply of vitamin C.

Milk is able to compensate for a lack of certain nutrients in a monotonous diet because of the great diversity of nutrients it contains and the high value of milk protein. It can therefore greatly improve the quality of the diet.

Products derived from milk contain these nutrients to a greater or lesser extent. Milk is especially desirable for vulnerable groups, for instance babies, toddlers, children and pregnant and nursing mothers. Always strive for a healthy, varied diet, which apart from milk also includes cereals, pulses, vegetables, fruits and if possible meat or fish. The various types of milk differ in various ways, including nutritional value.

COMPOSITION AND CHARACTERISTICS OF VARIOUS TYPES OF MILK

The composition of mother's milk and milk from cow, buffalo, goat, sheep, camel, donkey and lama is shown in table. The figures in table show that the composition of the milk of non-ruminants, *e.g.* mother's milk and mare milk, differs distinctly from the milk of ruminants (cow, goat, sheep, etc.). This may be partly explained by differences in the digestive system of the two groups.

Apart from the differences in cream formation there are other differences between the various kinds of milk. There is a lot of provitamin A (carotene) in cow milk, giving it its yellow colour, but not in buffalo, goat or sheep milk. In the milk of goats and sheep the carotenoids are already converted into the colourless vitamin A. This is why only cow milk is yellow in colour.

Buffalo milk curdles sooner than cow milk. Unless the preparation is adjusted, cheese made from buffalo milk will mature more slowly and have a drier consistency than cheese made from cow milk. Goat milk can have an unpleasant smell; this can be prevented by boiling the milk as soon as possible after milking.

Between some goats or breeds of goats there may be a difference in the taste of the milk. Cow milk accounts for 91% of the world's milk production. Buffalo, goat and sheep milk account for 5.9%, 1.6% and 1.7% respectively.

Table. Composition of various Types of Milk

Milk Source Energy (cal 100g)	Fat (%)	Protein (%)	Lactose (%)	Calcium (%)
Human 73 (mother's) milk	4.6	1.2	7.0	0.0
Friesian cow 62	3.5	3.3	4.6	01
Guernsey cow 75	4.7	3.2	4.7	0.1
Indian buffalo 100	7.5	3.8	4.9	0.2
Goat 71	4.5	3.3	4.4	0.1
Sheep 105	7.5	5.6	4.4	0.2
Mare 47	1.6	2.2	6.0	0.1
Donkey 46	1.5	2.1	6.2	0.1
Camel 70	4.2	3.7	4.1	?
Lama 65	3.2	3.9	5.3	?

Although there are enormous regional differences we can generally say that, if it is to be drunk, milk from cows or buffaloes is preferred to that from goats and sheep.

This is because of the more neutral flavour of cow milk and buffalo milk. Goat and sheep milk are, just like the milk of cows and buffaloes, popular for making cheese and soured milk products (especially sheep milk). Camel milk is usually drunk. Mother's milk is the most ideal food for a suckling infant. Nonetheless, many substitutes have been developed which find a ready demand. We shall pay more attention to infant nutrition in the next section.

INFANT NUTRITION

Mother's milk is best suited to the needs of a baby, and contains certain components that protect an infant against infectious diseases. All the nutrients a baby needs, except iron and vitamin C, are to be found in sufficient quantities in mother's milk. At birth, a baby has a store of iron in its liver, which it uses up during its first 6 months. Any kind of supplementary feeding is only necessary after 3 months, as the mother's milk then no longer supplies all the nutrients the infant needs. Fruit juice and mashed fruit provide additional vitamin C, which the infant then needs. Supplementary feeding of energy-giving foods is also desirable. Mixing small quantities of milk powder into mashed food can considerably improve the food's value (especially the value of its protein). It is advisable to continue breastfeeding as long as possible because mother's milk is often the only source of animal protein for a baby.

If the mother cannot breastfeed, does not have enough milk or dies, bottle feeding is a solution and the best substitute. However, in practice often too much water is added to the (artificial) baby food, which is usually bought in powder form. It becomes too watery and is therefore not nutritious enough. Moreover, artificial foods are costly and require good hygiene.

Dilution with water is often a cause of infection because the available water may be polluted. Water used for bottle feeding must first be boiled, but sterilising water by boiling uses a lot of fuel, which is often in short supply. By using a cup or a spoon it is easier to maintain the necessary hygiene rather than a bottle because they are easier to clean. Money might be better spent on essential necessities of life than on artificial infant food if the latter is not strictly necessary. If a baby cannot digest milk, you will be forced to use milk products, which do not contain lactose. This is the case with inherited lactose intolerance.

LACTOSE INTOLERANCE

Lactose intolerance means that the human body is almost, or entirely, unable to digest the milk sugar, lactose, which is present in milk because the body lacks the enzyme lactase. Lactase splits the lactose into glucose and galactose. The latter two mono-saccharides can easily be absorbed in the intestine.

Undigested lactose can be converted by the microbial flora in the intestine into lactic acid and gases. Consumption of larger quantities of milk thus causes flatulence, stomach cramps and diarrhoea. 'Lactose intolerance' is thus often called 'lactase deficiency'.

There are different forms of lactose intolerance among children:

- Congenital lactose intolerance. In this case, a baby cannot digest milk because the baby lacks the enzyme lactase, necessary for the breakdown of lactose into glucose and galactose.
- Lactose intolerance among children who are 2-5 years old. From the age of two years lactase activity in a child decreases and the child may have problems due to insufficient lactase by the time he or she is 4-5 years old. Consumption of small quantities of milk (one glass at a time)

usually does not cause any problem. It is also possible to prevent problems by eating fermented milk products, in which part of the milk sugar has been converted, such as cheese, yoghurt and buttermilk.

- Lactose intolerance as a result of intestinal disease and/or malnutrition, especially in babies and toddlers. The lactose activity is temporarily decreased making it necessary to use lactose-free milk products for a short time. Cheese and fermented products like yoghurt, in which milk sugar has been converted, are also suitable.

Apart from lactose intolerance, the use of milk also depends on other factors, which as suggested, now discuss.

MILK AND DAIRY PRODUCTS IN THE DIET

Diet refers to the way people feed themselves and the foodstuffs they use to do so. This is strongly influenced by people's traditions and religion, their economic position, their place in society and the possibilities offered by their natural surroundings. It is not surprising that each population group has its own diet. The use of milk and dairy products can also be looked at when examining the diet.

Here are a few examples of how the factors can influence the role and form of milk and dairy products in the diet:

- The cow is a sacred animal in India; therefore the rennet used in cheese making may not be taken from a calf's stomach.
- If milk or dairy products have to be bought, money is needed.
- In densely populated areas, people are forced to use all available land for crops that give a maximum yield, or crops that can be directly consumed by the people. This limits the land available for dairy farming.
- It may not be possible to keep cattle in certain regions, *e.g.* the humid tropics, due to the natural environment. For example, in humid areas of West Africa cattle cannot be kept because they are the host of the tsetse fly, which transmits sleeping sickness.

For these reasons, milk and dairy products in the diet can be of greater or lesser importance in one area or another. Economic and social situations are especially subject to change and dietary patterns change with them. New foodstuffs may be introduced to (partly) substitute others. Adoption of new products is often no easy matter; sometimes centuries-old traditions may have to change.

Also, taste and other characteristics such as texture are important in the acceptance of new kinds of food. Something that may be considered very tasty in one place, may not be appreciated elsewhere.

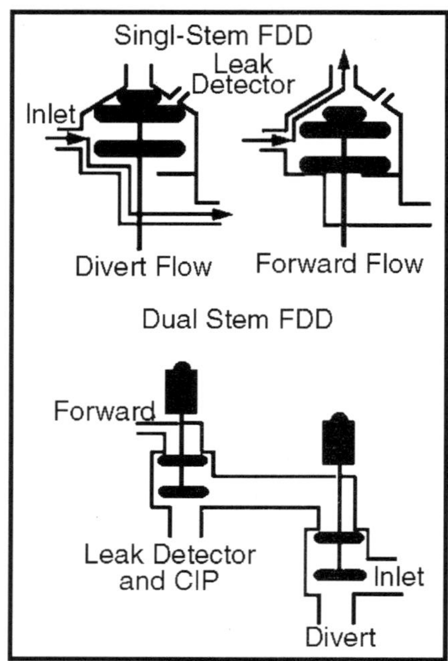

Fig. Camel Milk is a very Popular Food in Parts of Africa and the Middle East

RAW MILK TRANSPORT AND STORAGE

In developed countries, raw milk on the farm is usually cooled quickly and stored in refrigerated bulk tanks at <7 °C prior to collection. Collection by insulated tanker is often on alternate days, or sometimes less frequently, and therefore some of the milk in the tank could be 48 hours old at the time of collection. Temperature control is thus critical to minimise microbial growth, and tanker drivers are usually permitted to refuse milk stored at too high a temperature, or which has an abnormal appearance or odour.

Bacterial numbers in the milk may increase during transport, either as a result of contamination from inadequately cleaned tankers or from the growth of psychrotrophic organisms, particularly *Pseudomonas* spp.. Milk temperature and duration of the transport stage are therefore important factors. On arrival at the processing site, the milk is transferred to bulk storage tanks, or silos, prior to processing.

The milk may be stored in the silos for 2-3 days, and further growth of psychrotrophic bacteria is likely during this period. The degree of growth is dependent on the initial microbial load, and the storage time and temperature. Pseudomonads are the predominant organisms present in stored raw milk, with *Pseudomonas fluorescens*, *Pseudomonas fragi*, and *Pseudomonas lundensis* being commonly isolated, but Enterobacteriaceae, *Flavobacterium*, *Alcaligenes*, and Gram-positive species can also be found.

The growth of psychrotrophic bacteria may also be accompanied by the production of heat-stable, extracellular proteolytic and lipolytic enzymes. These enzymes are often capable of surviving pasteurisation and, in some cases, ultra high temperature processing, and they may subsequently cause spoilage in the processed milk. A number of techniques have been used to limit the growth of psychrotrophs during raw milk storage.

Thermisation

The most commonly used technique is to apply a mild heat treatment by heating to around 57-68 °C for 15-20 seconds and then cooling rapidly to <6°C. This reduces the psychrotrophic population significantly and can extend the storage life of the raw milk by several days. However, thermisation cannot eliminate vegetative pathogens, and is therefore not a reliable control for the hazard. For example, *L. monocytogenes* can survive the process and could then grow during chilled storage.

Deep Cooling

As the storage temperature is a key factor for the rate of growth of psychrotrophic spoilage organisms, storing milk at as Iow a temperature as possible can also extend the storage life significantly. Reducing the storage temperature from 6°C to 2°C has been shown to give a 2-day gain in storage life for milk of good microbiological quality.

Carbon Dioxide Addition

There has been some interest in extending the storage life of raw milk by the addition of carbon dioxide at a concentration of 20-30 mM. Three mechanisms are thought to be involved in carbon dioxide inhibition of microorganisms: the first is by the displacement of oxygen; the second is a lowering of the pH of the milk due to the dissolution of carbon dioxide and formation of carbonic acid, particularly for Gram-negative psychrotrophic aerobes; and the third is a direct effect on the metabolisms such as inhibiting the production of enzymes by these organisms.

It has also been suggested that the technique could be used to extend the shelf life of pasteurised milk, but concerns have been raised that the use of carbon dioxide addition could allow growth and toxin production by psychrotrophic *Clostridium botulinum*. However, recent work indicates that the risk of botulism is not increased by the use of this treatment. Following storage, the milk then undergoes further processing.

SEPARATION

If necessary, the milk is separated into skimmed milk, cream and sediment fractions, using centrifugal separators. The sediment may contain a comparatively high number of microorganisms and must be carefully discarded. The agitation involved may also break up clumps of bacteria, potentially producing an apparent increase in the number of colony-forming units. This process also allows the milk to be standardised to a specified fat content by adding back the correct quantity of cream.

HOMOGENISATION

The fat globules in milk can coalesce and form a cream layer. Homogenisation reduces the size of the milk fat globules by using a pump to force milk through a valve under pressure. The fat globules are then small enough to remain in suspension. This process has little microbiological effect, although clumps of bacterial cells may be broken up.

Homogenisers used for pasteurised milk may be linked to the pasteuriser, and run at raised temperature in order to minimise possible microbial contamination. UHT processed milks are homogenised in sterile conditions after heat treatment and before aseptic filling. Effective cleaning and sterilising of the homogeniser are then critical to product safety.

PASTEURISATION

Some form of heat process is commonly applied to milk to ensure microbiological safety, and to extend shelf life. In the UK, the most commonly used process is pasteurisation. Time-temperature requirements for pasteurisation vary between countries, and are often specified in legislation. In the UK, both low-temperature, long time and high-temperature, short time minimum processes are permitted.

However, in practice, the HTST process is now generally used. Recent concern about the possible survival of MAP in pasteurised milk has seen many dairies increase the length of the HTST process to 25 seconds.

Higher processes may also be applied to products with high fat and solids content. Plate heat exchangers are the most common method for milk pasteurisation, but it is essential that they are designed, constructed and operated in such a way as to minimise the possibility of recontamination of the pasteurised milk by raw milk.

Most commercial pasteurisers are fitted with sensors that continuously monitor the pasteurisation temperature, and are linked to automatic divert valves.

If the pasteurisation temperature falls below a specified value, the valve opens and diverts the under-processed milk away from the post pasteurisation section of the plant and the filling line, into a divert tank. The correct operation of these monitoring systems is critical and should be regularly checked. It is also essential that there are no cross-connections between the raw and pasteurised sides of the process, and this should include separate clean-inplace systems. It is also usual to maintain a higher pressure in the pasteurised milk to minimise the risk of cross contamination in the heat exchanger. Recontamination of this kind may have serious public health consequences. Accepted pasteurisation processes are designed to reduce the numbers of vegetative microbial pathogens to levels that are considered acceptable, although bacterial spores are not destroyed. Most of the potential psychrotrophic spoilage bacteria are also eliminated. However, certain heat-resistant mesophilic organisms, referred to as thermoduric, are able to survive pasteurisation.

Thermoduric species commonly isolated from pasteurised milk include *Micrococcus* spp., *Enterococcus faecium* and *Enterococcus faecalis*, *Bacillus subtilis*, *Bacillus cereus*, and certain lactobacilli. Psychrotrophic strains of these organisms may be able to grow slowly in the pasteurised milk at 5°C, and, if present initially in high numbers, could eventually cause spoilage. Effective cleaning of the cooling sections of pasteurisers is important to ensure that these organisms do not build up on surfaces.

UHT OR STERILISATION PROCESSES

Milk may also be subjected to more severe heat processes sufficient to achieve "commercial sterility". This may be done by batch heating in closed containers, or continuously with aseptic filling into sterile containers. Both conventional retort sterilisation and UHT processes must achieve a minimum Fo of 3 minutes to ensure product safety. These processes destroy all vegetative cells in the milk, and the majority of spores, although certain very heat-resistant spores may survive. This results in a long shelf life without the need for refrigeration, but also causes organoleptic changes in the milk, such as browning.

Conventional sterilisation processes involve heating the milk in thick-walled glass bottles, closed with a crimped metal cap, at about 120 °C for approximately 30 minutes. However, modern large-scale production methods often use an initial UHT treatment prior to filling the container, followed by retorting for a reduced time and then a rapid cooling process. This is said to give a product with improved organoleptic properties. UHT processes may be direct or indirect. Direct systems inject high-pressure steam directly into the milk to obtain the desired temperature, and then employ flash cooling under vacuum to remove the resulting excess water.

Indirect systems utilise heat exchangers and holding tubes. Direct systems are said to give better organoleptic properties, as the heating and cooling processes milk involves preserving milk by holding at a temperature of 140-150 °C for 1-2 seconds. Heat treatment is usually followed by aseptic filling into sterile cartons or other containers. The maintenance of sterility in filling is vital to prevent recontamination of the treated milk. As with pasteurised milk, it is also vital to ensure that raw milk cannot recontaminate the UHT-treated milk. Certain very heat-resistant spores of mesophilic bacilli, classified as *Bacillus sporothermodurans* are able to survive UHT processes and may subsequently grow in the final product. However, this organism has been shown not to be pathogenic and does not seem to cause any detectable changes to the product. Thermoduric *Bacillus stearothermophilus* are able to survive UHT processes and cause flat-sour spoilage.

PROCESSING AND ITS EFFECTS ON THE MICROFLORA

Examples of processes used to produce condensed, evaporated and dried milk products are shown in Figures. In each case, the initial steps of milk storage, transport, separation and standardisation are the same as those used for fresh milk. The first key stage is therefore the pasteurisation or pre-heating process. The time and temperature applied depends on the intended use of the product, but, as a minimum, will correspond to milk pasteurisation.

Much higher processes are used in some instances. For example, in large-scale continuous production of evaporated milks, temperatures as high as 121°C may be applied for several minutes to stabilise milk proteins.

Condensed milk is also subjected to processes more severe than milk pasteurisation, with the exact process being determined by the nature of the product required. This preheating helps to increase viscosity and improve other characteristics.

These processes may be less controlled than conventional pasteurisation, but it is important that a safe minimum process is always applied. As with fresh milk, most vegetative bacteria will be destroyed during heating, but some thermoduric types and bacterial endospores are likely to survive all but the most severe processes. The second common process in the manufacture of all types of concentrated milk products is the removal of water, usually by evaporation.

The most common type of evaporator used in the dairy industry is the falling film evaporator, which is both energy-efficient and readily controllable. It is common to link several evaporators together in series to form what is known as a 'multiple effect evaporator', with a common condenser and vacuum source.

The vapour produced in the first effect heats the second, and so on, producing a stepwise decrease in temperature from 70-80°C in the first effect, to about 40°C in the last. The vacuum, however, is greater in the lowest temperature effect, so that the milk flows from high to low temperature.

This process is very efficient and produces milk at the required concentration without a second pass. The temperatures within the lower temperature effects of the evaporator are low enough to permit the growth of thermophilic and some mesophilic spores, and quite high numbers may develop in the effects during prolonged production runs. Growth of certain thermophilic species has even been reported in milk at temperatures as high as 70°C. The growth of thermophiles must be controlled, by limiting the length of production runs, effective plant cleaning and sanitation, and ensuring that adequate standards of plant hygiene are applied.

CONCENTRATED MILK

Raw milk is given a heat treatment approximating that of pasteurisation. It is then concentrated at a low temperature, followed by standarisation, homogenisation and pasteurisation before packaging. Pasteurisation is usually done at 79.4°C for 25 sec; the product is stored at 10°C to prevent growth of thermoduric bacteria and any post-pasteurisation contaminants.

BULK CONDENSED MILK

In the manufacture of bulk condensed milk, the milk is first separated, if necessary, and standardisation of fat content is often carried out after concentration. Unless skimmed milk is used, homogenisation is also usually carried out at this stage.

A pre-heating process is then applied, using a continuous heater or a 'hot well'. Temperatures of 65.6–76.7°C are often used, but higher temperatures of 82.2-93.3°C for as much as 15 minutes may be applied to obtain a higher viscosity product and to impart other desirable characteristics.

These processes will greatly reduce the number of vegetative bacterial cells in the milk, but they are not under the same degree of control as conventional pasteurisation.

The heater or 'hot wells' can act as incubators for thermophilic bacteria, especially if the product is held in the lower temperature range, allowing bacterial numbers to build up; it is also possible that pathogens may survive some lower pre-heat temperatures. Therefore, a further HTST pasteurisation step should be applied to eliminate this hazard.

The preheated milk is then concentrated in a vacuum-pan or in a multiple-effect evaporator at a temperature range of 54.4-57.2°C After evaporation, the product is not sterile and will support rapid microbial growth.

It is also likely that post-process contamination could occur during standardisation or packaging. Therefore, effective post-process hygiene procedures, and rapid cooling to <5°C are necessary to achieve the required shelf life. The process is depicted in Figure.

Fig. Production of Condensed Milk

SWEETENED CONDENSED MILK

In the manufacture of sweetened condensed milk, the milk is pre-heated, and may also undergo superheating. Temperatures within the range 82-100°C for 10-30 minutes are used, and these processes are sufficient to destroy vegetative cells. For bulk, sweetened condensed milk, sugar can be added before concentration at varying levels depending on the end use. This material is not usually microbiologically stable and has a limited shelf life. In the production of sweetened condensed milk for retail sale, sugar is added during the later stages of evaporation. The sugar is normally added as a 65% solution, and the finished product then contains a high enough solute concentration to give an aw of 0.83-0.86. It is this reduced aw, rather than a heat process, that confers microbiological stability, and a long shelf life when packed in pre-sterilised cans or tubes. However, some osmophilic yeasts and moulds are able to grow in the product, and may cause spoilage.

Evaporation is carried out at a temperature of around 57.2°C, but can drop to 48-9°C late in the cycle. Partially cooled milk is seeded with very fine lactose crystals to force crystallisation. A high standard of hygiene during the later handling and filling stages is necessary to prevent contamination. A particular problem with sweetened condensed milk is its high viscosity, and 'sticky' nature, which makes cleaning processing equipment difficult. This high viscosity means that positivedisplacement, plunger-type fillers are needed to fill the product. These are complex and difficult to clean and may become heavily contaminated with micrococci and yeasts if not properly maintained.

EVAPORATED MILK

The manufacture of canned evaporated milk is very similar to that of bulk condensed milk, but the product is given a long shelf life by applying a heat process designed to give commercial sterility. Because of this, the milk has to be stabilised to prevent coagulation during processing and to minimise 'age thickening' during storage. Stabilisation is achieved by the addition of permitted salts, including phosphates, citrates and bicarbonates, which are used to maintain the pH of the milk at 6.6-6.7. Pre-heating is also important, and temperatures of 120-122 °C for several minutes are used to denature whey proteins. Only a few bacterial spores are likely to survive such a process. Condensation is usually performed at a temperature lower than 54.5°C. After evaporation, the milk is homogenised, cooled and stored. It is then standardised, and further stabilising salts may also be added at this point. It is important to ensure that cooling is rapid and sufficient to minimise any microbial growth during this procedure.

The milk is then filled into cans, hermetically sealed, and sterilised using batch retorts or continuous sterilisers. Processing at 115°C for 15-20 minutes, or 120°C for 10 minutes, has been traditionally applied, but recently there has been a move towards UHT processing followed by aseptic filling into pre-sterilised cans or cartons. Processing at 130°C for 30 seconds, to 150°C for less than one second may be used. Retorted canned milk is commercially sterile, and only extremely heat-resistant spores of organisms such as *Bacillus stearothermophilus* are likely to survive. These spores do not germinate unless the cans are then stored at high ambient temperatures. Aseptically filled evaporated milk may become recontaminated during filling, unless stringent hygiene procedures equivalent to those used in filling other UHT milk products are used. The process is depicted in Figure.

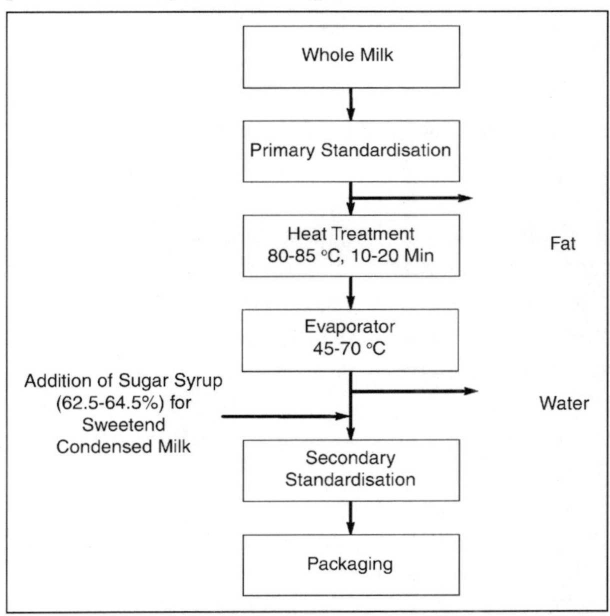

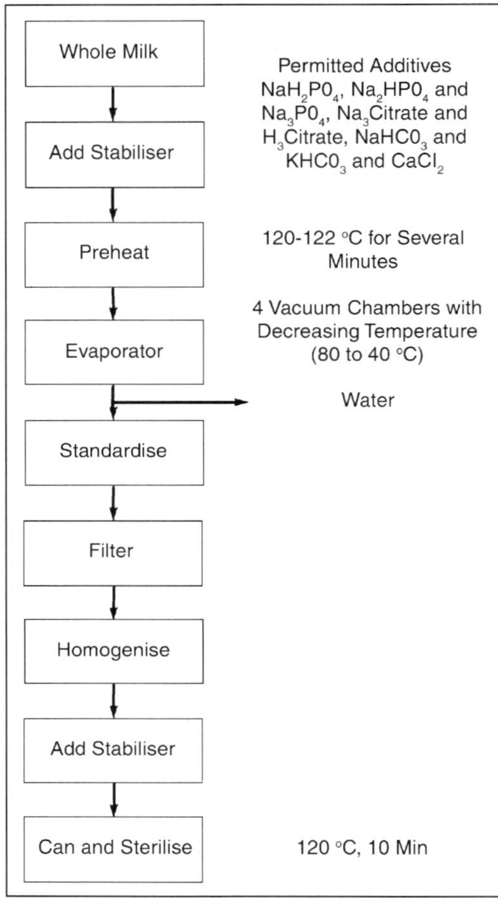

Fig. Production of Evaporated Milk

DRIED MILK

Drying is used, after evaporation, to produce a stable, powdered product with a residual moisture content of 2-5%. Most dried milk powders are now spraydried, and drum and roller dryers are little used. Spray drying is a more energy-efficient process and causes less heat damage to the product. Therefore, only spray drying will be considered further. Before evaporation and drying, the milk is standardised, if required, and heattreated.

Vegetative bacteria, including Enterobacteriaceae and *Listeria monocytogenes* have been shown to survive the drying process, and therefore all raw milk should receive a process at least equivalent to pasteurisation. Skimmed milk may also be subjected to low, medium and high heat processes to give varying degrees of protein denaturation as required. Typically, a low heat process will be 74°C for 30 seconds, a medium heat process 80-100°C for

1-2 minutes, and a high heat process may be equivalent to an ultra high temperature treatment. Milk for dried whole milk powders is heated at 85-95°C for several minutes.

UHT heating units result in products that have excellent microbiological quality, which is important when dried milk is to be used as an ingredient in baby food. The heat-treated milk is then evaporated to about 50% total solids before drying. As the drying process does not kill all vegetative bacterial cells, effective cleaning and hygiene procedures are required to ensure that the heat-treated milk does not become recontaminated. The current practice is to link the evaporation plant and dryer as a single integrated unit. Two systems of spray drying are used: jet or nozzle dryers, and rotary atomiser dryers.

Each dryer produces powders with specific characteristics. In the case of jet dryer systems, milk is fed to a high-pressure pump, then pumped under pressure to a series of jets or nozzles within the drying chamber. The impact of hot air on the milk causes the liquid to break up into fine droplets that form milk powder. In the rotary atomiser, the milk enters the drying chamber through slots or holes located at the periphery of a circular disc fixed to a rotating shaft.

8

Pasteurisation of Milk

Science and technology is undoubtedly escalating its claws in every area of nature, and inventions are being made towards advancement of civilization. The world today is not the same as it was decades ago, credit goes to technology growth. Nature has given us myriads of gifts, and technology has made variations in them, apparently for our benefits. There is a big controversy in the West, whether pasteurisation of milk is beneficial or detrimental to human health. Proponents of the process argue that it is to make milk and milk products safe for consumption by razing away injurious bacteria, and to improve keeping quality of raw milk.

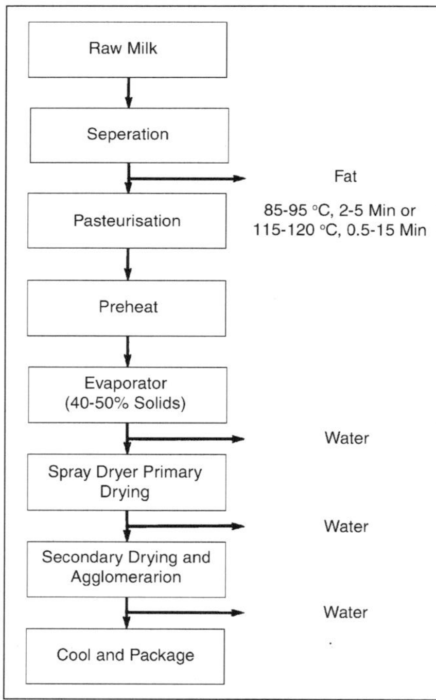

Fig. Drop of Milk.

Let us briefly know what pasteurization is. It is an artificial process used to destroy dangerous bacteria without substantially changing the composition or flavour of milk. The method also prevents rapid souring of milk. Under this process milk is heated to a temperature between 55 and 70°C for 30 minutes, and then is rapidly cooled and stored at a temperature below 10°C. Sometimes higher temperatures are applied for a shorter period of time.

The temperatures and time are determined by factors necessary to destroy bacteria (pathogens). The process was named after the French chemist Louis Pasteur. Pasteurization has been used to kill bacteria in milk including *Salmonella*, *Listeria*, *Campylobacter*, *E.coli*, and *Bracella*.

The problem is that in addition to the killing of bad bacteria, pasteurization kills good bacteria too. However, researches have been conducted and have proved that pasteurization diminishes the nutritional value of raw milk that is requisite for human health. Some of the effects of pasteu-risation are: the process destroys the enzyme phosphate present in raw milk. Phosphate is required to split and assimilate mineral salts in foods that are in the form of Phytates.

Protein, a valuable nutrient in raw milk, is greatly affected by the process. Its digestibility is reduced by 4 per cent and biological value is reduced by 17 per cent. The effect of the process on vitamins, vital nutrients in raw milk, is that the pasteurization process destroys about 38 per cent of vitamin B complex; and vitamin C is weakened or destroyed. Studies show that infants who are fed with pasteurised milk might develop scurvy. Not only protein and vitamins, but also mineral contents in raw milk are badly ruined by the process. After the process the calcium content is very much diminished.

The loss of soluble calcium in regards to infant's growth and development is a very important factor. It is not only the concern in children regarding bone formation and teeth, but also in adults about the calcium content in the blood. It also destroys 20 per cent per cent of iodine present in raw milk and causes constipation. After lot of research it is apparently clear that the process of pasteurization deteriorates quality of raw milk. Raw milk is in demand and is desirable by people who are becoming aware about the dangers of pasteurization to a product of daily consumption. Despite such scientific evidence in favour of raw milk and against pasteurised milk, and the fact that human beings have always lived on raw milk without facing any serious threats to life, an intriguing fact is that sale of raw milk has been made illegal in the US except in a few states.

People can only get pasteurised milk or organic milk. Organic milk is the milk that comes from the cows that are raised on organic food, no growth hormones or antibiotics are used and cattle graze on grass which hasn't been sprayed with pesticides. However, organic milk is around two and a half times the price of pasteurised milk, which becomes difficult for everyone to afford.

However, milk is such an important component in daily life, that despite so much of controversy about pasteurization and no good alternative for raw milk, people don't have choices but to have pasteurised milk in the US. The problem is not of the West anymore, but is rapidly transferring to countries like India too.

This is the issue created by technology and need to be raised in India before pasteurised milk becomes the choice of the Government of India. Pasteurization destroys most disease producing organisms and limits fermentation in milk, beer, and other liquids by partial or complete sterilization. Pasteurization does not destroy organisms that grow slowly or produce spores. While pasteurization destroys many microorganisms in milk, improper handling after pasteurization can recontaminate milk. Many dairy farms use a home-pasteurizing machine to pasteurise small amounts of milk for personal use. Raw milk can also be pasteurised on the stovetop.

Microwaving raw milk is not an effective means of pasteurization because of uneven heat distribution. Ultra-high temperature (UHT) processing destroys organisms more effectively and the milk is essentially sterilised and can be stored at room temperature for up to 8 weeks without any change in flavour. Milk, a natural liquid food, is one of our most nutritionally complete foods, adding high-quality protein, fat, milk sugar, essential minerals, and vitamins to our diet. However, milk contains bacteria that—when improperly handled—may create conditions where bacteria can multiply. Most of the bacteria in fresh milk from a healthy animal are either harmless or beneficial. But, rapid changes in the health of an animal, or the milk handler, or contaminants from polluted water, dirt, manure, vermin, air, cuts, and wounds can make raw milk potentially dangerous.

How do Microorganisms Enter the Milk Supply

Our environment contains an abundance of micro-organisms that find their way to the hair, udder, and teats of dairy cows and can move up the teat canal. Some of these germs cause an inflammatory disease of the udder known as mastitis while others enter the milk without causing any disease symptoms in the animal.

In addition, organisms can enter the milk supply during the milking process when equipment used in milking, transporting, and storing the raw milk is not properly cleaned and sanitised. All milk and milk products have the potential to transmit pathogenic (disease-causing) organisms to humans.

The nutritional components that make milk and milk products an important part of the human diet also support the growth of the organisms. Drinking raw milk causes foodborne illness, and dairy producers selling or giving raw milk to friends and relatives are putting them at risk.

COMMON PATHOGENS IN MILK

Illnesses from contaminated milk and milk products have occurred worldwide since cows have been milked. In the 1900s it was discovered that milk can transmit tuberculosis, brucellosis, diphtheria, scarlet fever, and Q-fever (a mild disease characterised by high fever, chills, and muscular pains) to humans.

Fortunately, the threat of these diseases and the incidence of outbreaks involving milk and milk products has been greatly reduced over the decades due to improved sanitary milk production practices and pasteurization.

Salmonella

Salmonellosis is the most common disease transmitted in raw milk. This organism is shed in the faces of cattle and picked up on the animals' hair or teats.

Many strains of *Salmonella* can cause foodborne illness in humans, and all strains exhibit the same symptoms such as gastroenteritis (vomiting and diarrhoea). Pasteurization destroys the *Salmonella* organism, and although pasteurised milk, powdered milk, and cheese have been implicated in salmonellosis outbreaks, in these cases, the pasteurised milk was contaminated during further processing.

Listeria monocytogenes

This widespread organism is found principally in soil. Listeriosis in humans may cause serious illness, and is especially dangerous to pregnant women, causing stillbirths or infant death soon after birth. Pasteurization inactivates *Listeria monocytogenes.*

Yersinia enterocolitica

This common organism has been found in many foods of animal origin including milk, cheese, and red. *Yersinia,* found in streams, lakes, and wells, spreads from the water to warm-blooded animals. The most common symptom of yersinosis is gastroenteritis and mimics the symptoms of appendicitis. *Yersinia enterocolitica* is destroyed by pasteurization.

Campylobacter jejuni

This organism, isolated in raw milk and meat, can cause mastitis in dairy cattle. It has also been isolated in the faces of many species including dogs, cats, rodents, cattle, sheep, swine, and poultry. Symptoms include vomiting, cramps, bloody diarrhoea, mild enteritis, or severe enterocolitis. Individuals who have recovered from the disease may suffer a relapse. *Campylobacter jejuni* is destroyed by pasteurization.

Staphylococcus aureus

It is a common cause of mastitis in dairy cattle and can enter the milk supply from sores on the teats of cows or from the hands and nasal discharges of dairy farmers and workers. The *Staphylococcus* organism produces an enterotoxin (toxins causing vomiting and diarrhea) in raw milk when it is held at temperatures above 50°F. Sufficient amounts of enterotoxin in foods can cause illness. The incidence of staphylococcal intoxication has been greatly reduced by pasteurization.

CHRYSEOBACTERIUM SPECIES IN MILK

Several Chryseobacterium species which include *C. balustinum*, *C. gleum* and *C. joostei* are associated with the spoilage of dairy products during cold storage. In milk they produce heat resistant proteolytic and lipolytic enzymes responsible for off-flavours. They also produce off-odours in pasteurised milk and cream, surface taint in butter, thinning in creamed rice and bitterness due to the production of phospholipase C. The production of these spoilage enzymes are dependent *inter alia* on temperature and growth phase. Consumers depend upon their senses of sight, smell, taste and touch to evaluate food quality.

A careful combination of microbiological-, sensory- and chemical analysis are required to determine the spoilage potential of a microorganism. Despite the importance of microorganisms in food spoilage, the definition and assessment of spoilage relies on sensory evaluation since neither the "total count" of *e.g.*, 107 cfu/ cm^2 nor the level of a specific spoilage organism *per se* can directly predict the sensory quality of a product. The growth and activity of spoilage microorganisms is mostly described and studied as a function of substrate base and of chemical and physical parametres such as temperature, pH, water activity and atmosphere.

The production of enzymes will always give valuable information on potential spoilage abilities while production of metabolites containing alcoholic compounds contribute to certain flavour compounds. Volatile and semi-volatile organic compounds present both in the matrix and the headspace aroma are largely responsible for the flavour qualities of the foods we eat. Flavour is considered to be the most important factor of any of the quality categories when comparing various dairy products. Dairy products provide a great sense of eating pleasure as a result of their characteristic flavour and smooth texture and they play an important role in a well-balanced diet. Because milk possesses a bland and soft flavour, the appearance of an objectionable off-flavour or off-odour is readily noticeable.

There is a delicate balance between many flavour compounds that contribute to a desirable milk flavour and if this balance is disturbed, off-flavours may occur. Flavobacteria can produce a putrid butter effect and a "sweaty feet" odour in skim milk. The aim of this study was to investigate the different growth and hydrolytic characteristics of the type species of the Chryseobacterium genus. These characteristics included the optimum growth temperature and growth capacity at different temperatures, pH values and NaCl concentrations. The proteolytic, lipolytic and phospholipase C production of the species was examined by using different microbiological culture media. The volatile compounds produced by the reference strains in milk, were examined by headspace gas chromatography. An odour description of milk samples, inoculated with the bacterial cultures were given by a semi-trained sensory panel of ten persons.

MATERIALS AND METHODS

Strains Investigated

Freeze-dried or agar slant cultures were reactivated in 10 ml Nutrient Broth and purity was checked by streaking on Nutrient Agar. Incubation was at 25°C for 24-48 h. The strains were maintained on Nutrient Agar slants and stored at 4°C.

Growth Characteristics

A battery of tests was performed to determine the different growth characteristics and to differentiate between Gram negative, yellow pigmented species, just as to the methods described by MacFaddin, Gerhardt *et al.*, Barrow and Feltham and Hugo *et al.*, unless indicated otherwise. All the tests were performed at 25°C, except where the growth of the reference strains at different temperatures, was examined on Nutrient Agar at 4°C, 25°C, 32°C, 37°C, 42°C and 55°C respectively.

Optimum Growth Temperature

The optimum growth temperature for four Chryseobacterium species was determined just as to the method described by Jooste *et al.* Four consecutive overnight cultures were prepared in tubes containing 10 ml each of Nutrient Broth. The Chryseobacterium species were incubated in a water-bath at each of the following temperatures: 20, 25, 30, 35, 40 and 45°C. Turbidity in the tubes, after 48 h, was read in a Milton Roy Spectronic 20D spectrophotometre at 660 nm using uninoculated Nutrient Broth as control. Optical density values were converted to Klett units.

Hydrolytic Activities

The proteolytic activity of the reference strains was determined by casein and gelatin hydrolysis, described by Harrigan and McCance. The lipolytic activity was determined using lecithinase production and the hydrolysis of olive oil and tributyrin. Phospholipase C production was determined using the method of Chrisope *et al.*

PRELIMINARY DETERMINATION OF VOLATILE COMPOUNDS IN MILK

Growth conditions were as follows: 100 ml samples of Clover, UHT process, fat free milk and full cream milk were measured aseptically into sterile Erlenmeyer flasks. Each milk sample was then inoculated with 1 ml of a 24 h old Nutrient Broth culture of *C. joostei*, *C. gleum* and *C. indologenes* respectively and incubated at 25°C for 72 h. Samples were prepared in duplicate. Analytical procedures were as follows: The volatile compounds were determined just as to human. Each of the inoculated milk samples and control samples were measured into a 10 ml headspace clear glass-vial and closed with crimpable headspace caps with silicone-PTFE seals.

After incubation for 30 minutes at 70°C, a 1 ml headspace sample was withdrawn using a gas-tight syringe and injected into a Varian Chrompack CP-3800 gas chromatograph equipped with a flame ionisation detector. A 30 m Zebron ZB-FFAP column, with a 25 μm film thickness and containing a nitroterephthalic acid modified polyethylene glycol liquid phase, with 40 to 250/ 260°C temperature limits was used. The temperature profile used was: 80°C for 2 min, increased by 5°C/ min to 250°C and held for 10 minutes. The total run time was 27 minutes. The carrier gas was hydrogen UHP. The column flow was 2.0 ml/min, the linear velocity 50.9 cm/ sec and the total flow 27.8 ml/ min. Identification of the unknown compounds was achieved by comparing their retention times to those of analytical grade standard compounds. An ethylacetate stock solution was prepared as standard and contained 200 mg.l-1 of each of propionic acid, valeric acid, 1-Heptanol, 1-Octanol, 1-Nonanol, 1-Decanol, 1-Pentanol, 1- Hexanol, 2-Octanone, 2-Butanone, 2-Heptanone, acetone, 4-Methyl-2- pentanol, 2-Methyl-1-butanol, ethyl acetate, methyl acetate, propyl acetate and n-buthanol.

Sensory Analysis of Milk

Ten candidates for the semi-trained sensory odour descriptive panel were screened on the basis of their ability to differentiate among odours and are all scientists employed by, or postgraduate students in the Food Science Division. The panel consisted of seven female and three male persons, ranging from 23 to 60 years of age. As part of the training process, the panellists performed a paired comparison test and were asked to identify the spoiled sample by doing a sniffing test. The spoiled sample consisted of 10 ml fat free milk which was inoculated with *Chryseobacterium joostei* and incubated at 25°C for 72 h. The control sample was uninoculated fat free milk. For the sniffing test, 10 ml samples of Clover, UHT Process, Fat Free and Full Cream milk in McCartney bottles were each inoculated with 0.1 ml of a 24 h old Nutrient Broth culture of C. joostei, *C. gleum* and *C. indologenes* respectively and incubated at 25°C for 72 h.

An uninoculated milk sample of each type of milk was included to serve as controls. Thirty minutes before sensory evaluation, the milk samples were incubated at 50°C to allow better assessment of the volatile fraction. The milk samples were presented in the same McCartney bottles, in which the inoculations were done. These bottles had tight fitting caps to prevent any unnecessary odour release, which could influence results. Samples were coded with three digit numbers and served randomly on white trays. The testing was done under red lights to mask any colour differences which occurred due to the characteristic yellow to orange pigmentation produced by Chryseobacterium species. The inoculated samples, along with two control samples, one each of the fatfree and full cream milk, were presented to the panel in individual booths. Samples were uncapped, sniffed and removed immediately so, as not to cause exhaustion.

When exhaustion did occur, panellists were instructed to smell their own skin or clothing. As part of the training, panellists were not allowed to wear any perfumed personal care products or smoke for 30 minutes prior to sniffing, as this could influence the results. Three spoilage levels were chosen to define the intensity of the spoilage: level 1, no spoilage noted; level 2, weak spoilage noted; level 3, strong spoilage noted. Panellists also described the type of odour, if any, produced by each of the three cultures. The samples were regarded as weakly spoiled and strongly spoiled when at least 50 per cent of the panellists rated it at level 2 and level 3, respectively. The sensory testing was done in individual booths at the sensory facility of the Food Science Division, Department of Microbial, Biochemical and Food Biotechnology, University of the Free State.

SPOILAGE OF FLUID MILK PRODUCTS

The shelf life of pasteurized milk can be affected by large numbers of somatic cells in raw milk. Increased somatic cell numbers are positively correlated with concentrations of plasmin, a heat-stable protease, and of lipoprotein lipase in freshly produced milk. Activities of these enzymes can supplement those of bacterial hydrolases, hence shortening the time to spoilage.

The major determinants of quantities of these enzymes in the milk supply are the initial cell numbers of psychrotrophic bacteria, their generation times, their abilities to produce specific enzymes, and the time and temperature at which the milk is stored before processing. Several conditions must exist for lipolyzed flavour to develop from residual lipases in processed dairy foods, that is, large numbers ($>10^6$ CFU/ml) of lipase producers, stability of the enzyme to the thermal process, long-term storage and favourable conditions of temperature, pH, and water activity.

SPOILAGE OF CHEESES

Factors that determine the rates of spoilage of cheeses are water activity, pH, salt to moisture ratio, temperature, characteristics of the lactic starter culture, types and viability of contaminating microorganisms, and characteristics and quantities of residual enzymes. With so many variables to affect deteriorative reactions, it is no surprise that cheeses vary widely in spoilage characteristics. Soft or unripened cheeses, which generally have the highest pH values, along with the lowest salt to moisture ratios, spoil most quickly.

In contrast, aged, ripened cheeses retain their desirable eating qualities for long periods because of their comparatively low pH, low water activity, and low redox potential. For fresh, raw milk pasta filata cheeses, Melilli determined that low initial salt and higher brining temperature (18 per centC) allowed for greater growth of coliforms, which caused gas formation in the cheese. Factors affecting the growth of the spoilage microorganisms, *Enterobacter agglomerans* and *Pseudomonas* spp. in cottage cheese, were higher pH and storage temperature of the cheese. Some of the spoilage microorganisms were able to grow at relatively low pH values when incubated at 7°C and were able to grow at pH 3. 6 when grown in media at 20°C. Rate of salt penetration into brined cheeses, types of starter cultures used, initial load of spores in the milk used for production, pH of the cheese, and ripening temperature affect the rate of butyric acid fermentation and gas production by *C. tyrobutyricum*.

Fungal growth in packaged cheeses was found to be most significantly affected by the concentration of CO2 in the package and the water activity of the cheese. Cheddar cheese exhibiting yeast spoilage had a high moisture level (39.1 per cent) and a low salt in the moisture-phase value (3.95 per cent). Roostita and Fleet determined that the properties of yeasts that affected the spoilage rate of Camembert and blue-veined cheeses were the abilities to ferment/assimilate lactose, produce extracellular lipolytic and proteolytic enzymes, utilize lactic and citric acid, and grow at 10°C.

CONTAMINATION OF RAW MILK

The highly nutritious nature of dairy products makes them especially good media for the growth of microorganisms. Milk contains abundant water and nutrients and has a nearly neutral pH. The major sugar, lactose, is not utilized by many types of bacteria, and the proteins and lipids must be broken down by enzymes to allow sustained microbial growth. In order to understand the source of many of the spoilage microflora of dairy products, it is best to discuss how milk can first become contaminated, via the conditions of production and processing.

The mammary glands of many very young cows yield no bacteria in aseptically collected milk samples, but as numbers of milkings increase, so do the chances of isolating bacteria in milk drawn aseptically from the teats. The stresses placed on the cow's teats and mammary glands by the very large amounts of milk produced and the actions of the milking machine cause teat canals to become more open and teat ends to become misshapen as time passes. These stresses may open the teat canal for the entry of bacteria capable of infecting the glands.

Environmental contaminants represent a significant percentage of spoilage microflora. They are ubiquitous in the environment from which they contaminate the cow, equipment, water, and milkers' hands. Since milking machines exert about 38 cm (15 in.) of vacuum on the teats during milking, and since air often leaks into the system, bacteria on the surfaces of the cow or in water retained from premilking preparation can be drawn into the milk.

Also, when inflation clusters drop to the floor, they pick up microorganisms that can be drawn into the milk. The pumping or agitation of milk supplies the oxygen needed by aerobes for growth and breaks chains and clumps of bacteria. Single cells, having less competition than those in colonies, have the opportunity for more rapid multiplication.

Bacteria recontaminating pasteurized milk originate primarily from water and air in the filling equipment or immediate surroundings and can be resident for prolonged periods of time. In a study performed in Norway and Sweden, Ternstrom, Lindberg, and Molin investigated nine dairy plants and found that five taxa of psychrotrophic *Pseudomonas* spp. were involved in the spoilage of raw and pasteurized milk and that the same strains were recovered from both the raw and pasteurized milk, suggesting that recontamination originated from the raw milk.

Additionally, the investigators found that *Bacillus* spp. (mainly *B. cereus* and *B. polymyxa*) were responsible for spoilage in 77 per cent of the samples that had been spoiled by Gram-positive bacteria. The spoilage *Bacillus* spp. grew fermentatively, and most were able to denitrify the milk, which has implications for cheeses that contain added nitrate/nitrites for protection against clostridia. Sporeforming bacteria are abundant in dust, dairy feed concentrates, and forages; therefore, they are often present on the skin and hair of cattle from which they can enter milk. The presence of sporeformers such as *C. butyricum* in milk has been traced to contaminated silage.

CONTAMINATION OF DAIRY PRODUCTS

Washed curd types of cheeses are especially susceptible to growth of coliforms, so great care must be taken to monitor the quality of water used in these processes. A high incidence of contamination of brine-salted cheeses by yeasts results from their presence in the brines. Many mold species are particularly well adapted to the cheese-making environment and can be difficult to eradicate from a production facility. Fungi causing a "thread mold" defect in Cheddar cheeses were found in the cheese factory environment, on cheese-making equipment, in air, and in curd and whey. In a study of cheese-making facilities in Denmark, *Penicillium commune* persisted in the cheese coating and unpacking areas over a 7-year period. Ascospores of *B. nivea* and other heat-resistant species shown to be able to survive pasteurization, such as *Talaromyces avellaneus*, *Neo-sartorya fischeri* var. *spinosa*, and *Eupenicillium brefeldianum*, have also been found in raw milk.

A major cause of failure of processing and packaging systems is the development of biofilms on equipment surfaces. These communities of microorganisms develop when nutrients and water remain on surfaces between times of cleaning and reuse. Bacteria in biofilms (sessile form) are more resistant to chemical sanitizers than are the same bacteria in suspension (planktonic form). Chemical sanitizers may be rendered ineffective by biofilms leaving viable bacteria to be dislodged into the milk product.

PREVENTION OF SPOILAGE IN CULTURED DAIRY PRODUCTS

Cultured products such as buttermilk and sour cream depend on a combination of lactic acid producers, the lactococci, and the leuconostocs to produce the desired flavour profile. Imbalance of the culture, improper temperature or ripening time, infection of the culture with bacteriophage, presence of inhibitors, and/or microbial contamination can lead to an unsatisfactory product. A buttery flavour note is produced by *Leuconostoc mesenteroides* subsp. *cremoris*. This bacterium converts acetaldehyde to diacetyl, thus reducing the "green" or yogurt-like flavour. A diacetyl to acetaldehyde ratio of 4:1 is desirable, whereas the green flavour is present when the ratio is 3:1 or less.

Proteolysis by the lactococci is necessary to afford growth of the *Leuconostoc* culture, and citrate is needed as substrate for diacetyl production Although cooking of the curd destroys virtually all bacteria capable of spoiling cottage cheese, washing and handling of the curd after cooking can introduce substantial numbers of spoilage microorganisms. It is desirable to acidify alkaline waters for washing cottage cheese curd to prevent solubilization of surfaces of the curd.

However, more pseudomonads can be adsorbed onto cottage cheese curd from wash water when adjusted to pH5 (40–45 per cent) rather than adjusted to pH7 (20–30 per cent). Flushing packages of cottage cheese or sour cream with CO_2 or N_2 suppressed the growth of psychrotrophic bacteria, yeasts, and molds for up to 112 days, but a slight bitterness can occur in cottage cheese after 73 days of storage. Cheesemakers can use the addition of high numbers of lactic acid bacteria to raw milk during storage to reduce the rate of growth of psychrotrophic microbes. For fresh, raw milk, brined cheeses, gassing defects can be reduced by presalting the curd prior to brining and reducing the brine temperature to $<12°C$. Pasteurization will eliminate the risk from most psychrotrophic microbes, coliforms, leuconostocs, and many lactobacilli, so cheeses made from pasteurized milk have a low risk of gassiness produced by these microorganisms.

Most bacterial cells, including spores, can be removed from milk by centrifugation at about 9,000*g*. The process, known as bactofugation, removes about 3 per cent of the milk, called bactofugate. Kosikowski and Mistry invented and patented a process for recovering this bactofugate which is heated at 135°C for 3–4 s, then added back to the cheese milk. The process can reduce the population of butyric acid-producing spores by 98 per cent. Spore-forming bacterial growth and subsequent gas production in aged, ripened cheeses can be minimized with a salt to moisture content of ≥3.0 per cent.

Other potential inhibitors of butyric acid fermentation and gas production in cheese are the addition of nitrate, addition of lysozyme cold storage of cheese prior to ripening, direct salt addition to the cheese curd, addition of hydrogen peroxide, or use of starter cultures that form nisin or other antimicrobials. The most popular mold inhibitors used on cheeses are sorbates and natamycin. Sorbates tend to diffuse into the cheese, thereby modifying flavour and decreasing their concentration, whereas very little natamycin diffuses.

Electron beam irradiation, studied by Blank, Shamsuzzaman, and Sohal for mold decontamination of Cheddar cheese, can reduce initial populations of *Aspergillus ochraceus* and *Penicillium cyclopium* by 90 per cent with average doses of 0.21 and 0.42 kGy, respectively. Since nearly all mold spores are killed by pasteurization practices that limit recontamination and growth, although difficult, are vital in prevention of moldy cheeses.

Modified atmosphere packaging (MAP) of cheeses can retard or prevent the growth of molds, and optimum MAP conditions for different types of cheeses were described by Nielsen and Haasum. For processed cheeses containing no active lactic acid starter bacteria, low O_2 and high CO_2 atmospheres were optimum; for cheeses containing active starter cultures, atmospheres containing low O_2 and controlled CO_2 using a permeable film provided the best results. For mold-ripened cheeses requiring the activity of the fungi to maintain good quality, normal O_2 and high, but controlled, CO_2 atmospheres were best.

In Italian soft cheeses such as Stracchino, vacuum packaging decreased the growth of yeasts, resulting in a shelf life extension of >28 days. Processing times and temperatures used in the manufacture of cream cheese and pasteurized process cheese are able to eliminate most spoilage microorganisms from these products. However, the benefit of the presence of competitive microflora is also lost. It is very important to limit the potential for recontamination, as products that do not contain antimycotics can readily support the growth of yeasts and molds. Sorbates can be added; however, their use in cream cheese is limited to amounts that will not affect the delicate flavour.

PREVENTION OF SPOILAGE IN OTHER DAIRY PRODUCTS

The high salt concentration in the serum-in-lipid emulsion of butter limits the growth of contaminating bacteria to the small amount of nutrients trapped within the droplets that contain the microbes. However, psychrotrophic bacteria can grow and produce lipases in refrigerated salted butter if the moisture and salt are not evenly distributed. When used in the bulk form, concentrated (condensed) milk must be kept refrigerated until used.

It can be preserved by addition of about 44 per cent sucrose and/or glucose to lower the water activity below that at which viable spores will germinate (a_w 0.95). Lactose, which constitutes about 53 per cent of the non-fat milk solids, contributes to the lowered water activity. When canned as evaporated milk or sweetened condensed milk, these products are commercially sterilized in the cans, and spoilage seldom occurs. Microbial growth and enzyme activity are prevented by freezing. Therefore, microbial degradation of frozen desserts occurs only in the ingredients used or in the mixes prior to freezing.

9

Animal's Waste Management in Dairy Farms

Traditional small dairy farms effectively manage manure by applying it to their fields as a crop fertilizer. Unfortunately, Indian farm policy and economics are causing the demise of those small, diversified farms and the rise of industrial factory dairies that cram together thousands of cows who make millions of gallons of manure. One 2,500-cow dairy produces as much waste as a city with 400,000 residents.

Factory dairy operators often attempt to deal with the high volume of manure, as well as waste from the milking operation, by using water to flush it out of buildings and into multimillion-gallon lagoons. Then, the mixture is sprayed or spread onto fields to fertilize them. But unlike the relatively small amount of cow manure coming from traditional farms, these loads of industrial dairy waste are too much for the land to absorb and filter. Instead, the manure pollutes the air, water, and soil. It also causes human health problems.

Storing millions of gallons of manure and other waste in one place emits dust particles and hundreds of different volatile gases, including ammonia, carbon dioxide and methane. In fact, two California industrial dairy workers died in 2001 from methane overexposure while they were working in a 30-foot deep manure pit. The dairy foreman and the manager were indicted for involuntary manslaughter over the deaths in February 2003. In addition to methane, the lagoons give off hydrogen sulfide, which can cause brain damage. It threatens the safety of workers and people living nearby.

Manure from factory dairies also poisons waterways and wells when, a lagoon wall collapses, an operator fails to handle pumps properly, or the waste spills out of lagoons during heavy rains.

Manure contains nitrates and phosphorus. When too many of these nutrients enter waterways, they prompt the rapid growth of algae and other aquatic plants that consume much of the oxygen in the water. This smothers fish and other animals and causes serious taste and odor problems in drinking water.

Each state issues permits that allow factory dairies and other confined animal feeding operations to operate. Regulators issue relatively few fines and rarely shut down the dairies because of spills or other pollution problems. In countries where cows are grazed outside year-round there is little waste disposal to deal with. The most concentrated waste is at the milking shed where the animal waste is liquefied (during the water-washing process) and allowed to flow by gravity, or pumped, into composting ponds with anaerobic bacteria to consume the solids. The processed water and nutrients are then pumped back onto the pasture as irrigation and fertilizer. Surplus animals are slaughtered for processed meat and other rendered products.

In the associated milk processing factories most of the waste is washing water that is treated, usually by composting, and returned to waterways. This is much different from half a century ago when the main products were butter, cheese and casein, and the rest of the milk had to be disposed of as waste (sometimes as animal feed).

In areas where cows are housed all year round the waste problem is difficult because of the amount of feed that is bought in and the amount of bedding material that also has to be removed and composted. The size of the problem can be understood by standing downwind of the barns where such dairying goes on.

In many cases modern farms have very large quantities of milk to be transported to a factory for processing. If anything goes wrong with the milking, transport or processing facilities it can be a major disaster trying to dispose of enormous quantities of milk. If a road tanker overturns on a road the rescue crew is looking at accommodating the spill of 10 to 20 thousand gallons of milk without allowing any into the waterways. A derailed rail tanker-train may involve 10 times that amount. Without refrigeration, milk is a fragile commodity and it is very damaging to the environment in its raw state. A widespread electrical power blackout is another disaster for the dairy industry because both milking and processing facilities are affected.

In dairy-intensive areas the simplest way of disposing of large quantities of milk has been to dig a large hole in the ground and allow the clay to filter the milk solids as it soaks away. This is not very satisfactory.

DAIRY MANURE MANAGEMENT

Dairy manure management can be classified into three systems — solid, slurry or lagoon — depending on the collection, transportation and distribution of manure on the fields. These systems are dependent on the amount of bedding and water dilution used by a specific operation. Many dairies use more than one system. Manure from dry cows and replacement heifers may be handled as a solid, whereas manure from free-stall barns, holding pens and milking parlors may be handled as a liquid, or solids may be separated from the liquid wastes so it can be used for other purposes. Solid systems are commonly used in smaller operations (less than 100 cows) with bedded loafing barns or stanchion stalls. These systems minimize the volume of manure that is handled; however, a separate facility is required for liquid milking centre manure.

Slurry systems maximize recovery of plant nutrients from manure and are often used where geologic conditions are unsuitable for a lagoon system. These systems increase the volume of manure to be handled because of the higher water content, but allow the manure to be handled as a fluid. Many dairies favour lagoon systems because costs are lower than for other systems. Lagoons are applicable where flushing is desired and where a significant amount of lot runoff must be contained.

To minimize pollution potential and maximize nutrient recovery, incorporate livestock manure into the soil as soon as possible after application. To prevent applying an excess of plant nutrients (N, P and K), keep long-term records on the amount applied to each field. Make an analysis to determine the nutrient content of the manure and keep an estimate of the amount of manure applied. The amount of land required for manure disposal is usually based on 100 pounds of nitrogen used annually for plant production per acre.

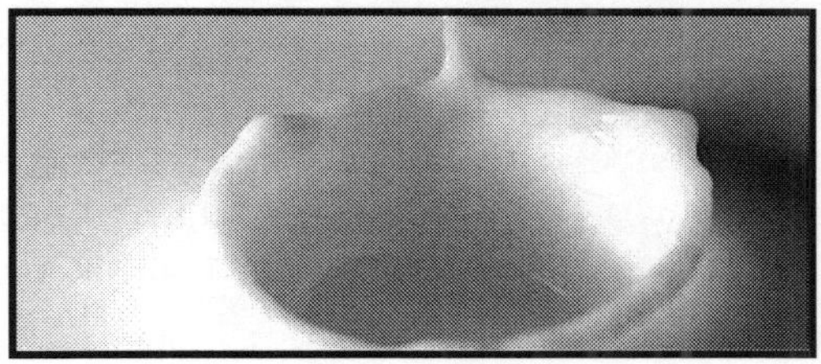

Fig. Dairy Waste Management

Consult local health and regulatory authorities and have all plans approved before constructing any manure handling system. A dairy operation with 700 or more cows is required by state law to obtain a permit. Any livestock manure management system, regardless of size, must be designed and operated in a manner that will not pollute surface or groundwater.

Local Extension centre or Natural Resources Conservation Service (NRCS) office may be contacted for information about permits and assistance in developing manure management systems.

SOLID WASTE SYSTEMS

Manure with a 75 to 80 percent moisture content can usually be handled as a solid. Manure at this moisture content has a consistency of peanut butter. Dairy manure, as excreted, contains about 88 percent water. So, to be handled as a solid, dairy manure must have the liquids drained off or bedding added. Twelve pounds of bedding per 100 pounds of fresh manure (about 4 pounds of dry straw per cow per day) is needed to permit handling dairy manure as a solid.

Scraping Systems

Manure from bedded housing usually has to be handled as a solid. As a semisolid, manure can be scraped into a holding area where liquids are allowed to drain off. Scraping can be done mechanically by alley or by gutter scrapers pulled by cables or chains. Small tractors with front- or rear-mounted blades are commonly used to scrape manure, and tractor scrapers function better in cold weather and tend to be more dependable than gutter scrapers.

Loaders

Front-end tractor loaders can remove manure from large areas with high clearances. Traction and stability can be a problem, however; wide front axles can improve stability and weights on the rear wheels or front-wheel-assist can improve traction. Skid steer loaders can work in cramped quarters and, in general, lessen time and hand labour expended.

Manure Storage

Storage is required with most solid handling systems because of weather or limited available fields. A 60-day minimum storage capacity is recommended. The storage area should have all-weather access with surface runoff water diverted from the storage area. Any draining liquid should be directed to a soil-plant filter, lagoon or holding pond. A roof can eliminate the need for collecting and storing runoff from the storage area.

Manure Spreaders

Most solid manure manure spreaders have moving aprons to move the load rearward to beaters. Flail-type spreaders have chains attached to a shaft that revolves at high speed to flail the manure out of the tank. These spreaders are also suited for semisolid manure.

SLURRY WASTE SYSTEMS

Manure with 90 to 96 percent moisture content can usually be handled as a fluid but may require special pumps. Manure at this moisture content has the consistency of malt. Manure with 96 to 98 percent water content can be handled with ordinary pumps and flushing equipment if excessive straw or fibrous material is not present. For conventional pumping, two gallons of water must be added to dilute a gallon of fresh manure to 96 percent. Of course, this greatly increases the volume of manure to be stored and transported.

Flushing is not well suited to a slurry manure management system. The slurry system is designed to minimize dilution because it uses tank wagons for transportation and application. Therefore, hauling extra water with the manure is not profitable. Assume a 1400-pound cow produces 14 gallons of undiluted manure per day. Hauling her yearly manure output will require 1.7 trips to the field with a 3,000-gallon tank wagon. If manure is diluted to 4 percent solids, 5.1 trips per cow per year will be required.

Slurry Manure Storage

Slurry can be stored in earthen basins, below-ground and above-ground concrete tanks and in lined and unlined metal tanks. Slurry storage areas are usually designed for a 120-day minimum storage capacity. In some locations, soil type can prohibit the use of earthen basins or require the use of a membrane-type liner. In these locations, concrete or steel manure storage tanks may be used. Depending on the location of buildings and lots, it may be possible to gravity drain or scrape manure into the storage structure. In-ground tanks must be designed to resist lateral forces from the soil when the tank is empty. Outside drainage may be necessary to prevent the tank from floating when it is empty and the soil is saturated. Most above-ground tanks have the added expense of a pump to transfer manure in and out of the facility.

Earth Basins

Earth basins are usually partially below the original grade and partially above. The above-ground berms divert surface runoff water from the basin. Check with the Division of Geology and Land Survey for suitable sites, soil suitability and construction details to avoid seepage polluting surface water or groundwater. Earthen basins must be constructed with a clay seal (bottom and sides) to prevent leakage. In some cases, an artificial liner may be required. If soils have a high potential for collapsing, MDNR will not approve an earthen manure storage structure. In these cases, an alternate site must be selected, or a concrete or steel tank used for storage.

Scrape to Storage

Scraping into below-ground storage areas can reduce equipment investment and maintenance costs. Any system that involves uncovered concrete lots, alleys or exercise areas increases the amount of runoff water that is hauled. Above-ground tanks require scraping into a collection pit and then pumping.

Pumps

Pumps are frequently used to move manure into tanks and to move manure from tanks into tank wagons. High capacity (1,000 to 3,000 gpm) chopper pumps are used to move dairy manure from storage, with extra capacity to mix solids and liquids.

Tank Wagons

Tractor-drawn tank wagons have capacities from 1,000 to 3,000 gallons. For long hauls and/or large operations, a tank truck may be more economical.

Tank wagons can either spread the liquid on the soil or be equipped with knives to inject the manure into the soil. Injection reduces odors and nitrogen loss but requires considerable power to pull the tank wagon and injectors at a reasonable speed. Injection can not be done when the soil is frozen or if it is too rocky. Likewise, spreading manure on frozen soil could pollute surface water if it rains or snows while the ground is still frozen. Slurry manure systems require more land for application than lagoon systems because more nitrogen is retained.

LAGOON WASTE SYSTEMS

Lagoon systems handle highly diluted manure (96 percent or more water) that can be pumped through irrigation systems. Most lagoon effluent is more than 99 percent water, so it has the consistency of water. Flushing is the preferred method of manure collection and transportation for many dairies because it eliminates the labour required for scraping and moving manure to storage. Lagoons complement flushing systems because they provide storage at relatively low cost and usually do not cause unacceptable odor problems if judiciously located. Lagoons can also provide a source of flush water.

The anaerobic breakdown of manure usually results in acceptable odor levels when applied by an irrigation system.

Approximately 90 percent of the input nitrogen is lost or unavailable in an unagitated lagoon, but this can be an advantage if limited land is available for manure disposal. For large operations, lagoons provide the least-cost alternative on a per-cow basis.

Consideration should be given to separating fibrous material (free-stall bedding, wasted hay, silage, etc.) from the manure flow going to the lagoon. Experience has shown that excessive amounts of this material can build up as surface crust and increase the rate of sludge accumulation in the lagoon, thus decreasing its long-term effectiveness as a manure storage and treatment facility.

Lagoons are usually designed for a 365-day storage capacity with all manure going into the lagoon. This includes dry cow manure, but not calf and replacement heifer manure. Also note that lagoons are not allowed by MDNR in areas where soil collapse potential is high.

Lagoons are usually formed by excavation and above-ground berms. The berms divert surface runoff water from the lagoon, but a diversion channel may be required.

For a "DNR approved system," a clay seal (compacted by a sheepsfoot roller) must be on the bottom and sides to meet NRCS specifications. Also, lagoons must be filled with fresh water to the minimum operating level and the berms seeded and mulched prior to operation.

Pumps

Conventional irrigation pumps are adequate for pumping lagoons if the intake is floated about 1 to 2 feet below the surface. The floating intake prevents surface debris interference and reduces the handling of sludge. Centrifugal power-take-off irrigation pumps are a practical alternative for irrigating from lagoons. Small engine-mounted low-lift pumps can also provide adequate capacity and pressure to pump a small lagoon.

A complete manure management system includes a way of spreading manure in a pollution-free manner. It is important that all irrigation systems be monitored while effluent is being applied so no runoff occurs at the application site. The following are ways to spread manure:

Surface Irrigation Systems

Gated pipe can provide a low-cost distribution system for pumping small lagoons, especially where uniform application is not critical. Small systems can use a "homemade" gated pipe made from polyethylene pipe (e.g., using 2-inch diameter pipe for 50 gpm systems) with holes in the pipe at 30- to 40-inch intervals. The holes must be small enough that flow is distributed out of all of the holes, but large enough that they don't become clogged by small solids. These systems are feasible for small lagoons handling only milkhouse manure but have been used for larger lagoons. For larger systems, commercial gated pipe with adjustable gates is an alternative.

Sprinkler Irrigation Systems

For irregular rolling land, use a sprinkler irrigation system instead of gated pipe. Various sprinkler systems are suitable; the choice depends on the desire to save labour or money.

Stationary Hand-carried Systems

For small operations, one or more small sprinklers fed by 3- to 6-inch portable pipe is adequate if time and labour are available. For even smaller operations, a 1 DA 2- to 1-inch nozzle sprinkler(s), covering 1 DA 2 to 2 acres per set, can be used. For larger operations, a high volume gun sprinkler with a 1- to 2-inch nozzle, covering 2 1 DA 2 to 6 acres per set, can be used to reduce labour requirement.

Moving Sprinkler Systems

To reduce labour and to accommodate larger operations, a traveling gun or a centre-pivot irrigator may be more cost-effective than hand-moved sprinklers. Different models of the small traveling guns are now available. Larger traveling guns, used for crop irrigation, can be used if crops are irrigated conventionally or if rental units are available.

DAIRY MANURE MANAGEMENT

Concerns regarding nutrient losses from the manure of large dairy herds to groundwater or surface runoff have been extremely acute in Florida. The most widely publicized concerns have been with phosphorus (P) contamination of Lake Okeechobee, probably washing off the farms in surface runoff during the summer rainy season. Also of concern are nitrogen (N) losses in the form of nitrate into the groundwater through the deep sandy soils of the Suwannee River basin.

Florida is not unique. All states are starting to monitor farms where large numbers of food-producing animals are maintained on small acreage. But regardless of farm size, maintenance of the environment and efficient utilization of nutrients should be a priority. Nitrogen (N) and phosphorus (P) are the primary nutrients of concern. Manure N can be converted to nitrates in the soil and potentially leach to groundwater.

Manure P can potentially move with surface runoff, thus stimulating algae and aquatic plant growth leading to eutrophication of surface waters. Potassium (K) has not been of as much concern environmentally as N and P, but it is of equal concern in fertilizer management strategies. Also, K is important in assessing equivalent commercial fertilizer value of manure. Manure nutrients and decaying organic matter are natural components of the environment that ultimately contribute to the production of more plant and animal tissue. Thus, concern about manure leakage is no different than concern about commercial fertilizer use. Manure is in fact a resource, based largely on its equivalence with commercial fertilizer, and should not be called a waste when it is recycled through new plant growth.

To use the resource value of manure effectively and to avoid excessive concentration of these nutrients at inappropriate points, it is helpful to budget nutrient flow through the total dairy farm system. If there is a problem concentration at some point, then corrective measures can be taken which are environmentally accountable.

This publication is designed to give background information and serve as a guide to dairy farmers and planners to help develop answers to these questions for an individual dairy farm.

Abbreviations: N = nitrogen, P = phosphorus, K = potassium, CP = crude protein, DM = dry matter, DMI = DM intake, Mcal = megacalorie of energy, NE_L = net energy for lactation, OM = organic matter, TDN = total digestible nutrients.

Individual Nutrientsis Excreted by Dairy Cow

Historically, the nutrient excretion standards most often used in the design of manure management systems were those of the American Society of Agricultural Engineers. These standards are based on body weight of cows, however, they do not account for the large variation among dairies in nutrient excretion levels. The variation is caused by differing voluntary feed intake, differing supplemental levels, and differing amounts of nutrients harvested in the milk.

Recent University of Florida experiments showed that N and P excretions by dairy cows varied dramatically with level of N and P intakes. These data confirmed that excretion estimates for individual farms are best based on dietary intake of a nutrient minus the amount secreted into milk. Most dairy farms have accurate estimates of feed and milk and, thus, this method of predicting total excretion of minerals by dairy cows is used in this publication. The following milk composition, typical of Holsteins, was used along with pounds of milk to determine recovery of fed nutrients in milk:

- Protein3.30 per cent (N content.512 per cent)
- Phosphorus (P)0.10 per cent
- Calcium (Ca)0.12 per cent
- Potassium (K)0.15 per cent
- Magnesium (Mg)0.01 per cent
- Sodium (Na)0.05 per cent
- Chlorine (Cl)0.11 per cent

Dairy farmers have considerable control of mineral excretion because they control the mineral contents in the diets they feed. Feeding adequate P is important for animal health and performance. However.40 per cent of total diet dry matter is near the year-round estimated requirements for lactating cows, more is recommended in early lactation.

Although these data do not lead us to recommend lowering feeding levels for P below standard feeding recommendations (NRC), the data point out that future committees that develop NRC feeding standards need to review recommended feeding levels for P, N, and environmentally sensitive minerals. The objective is to minimize excretion of these nutrients and still maintain optimum animal performance.

One is for cows consuming diets formulated to supply NRC crude protein standards which require more crude protein to assure that true protein needs are met (NRC, high). The other (NRC, low) minimizes dietary N by providing minimal nonprotein N and ruminally degradable protein for optimum rumen microbial fermentation and provides for remaining animal requirements with ruminally undegraded protein. Numeric estimates of yearly N excreted by high-producing 1400 lb cows were 273 lbs N per cow per year when fed according to the NRC crude protein standards and 234 lbs N per cow per year when diet protein was formulated for minimum needs for undegraded and degraded protein.

Diets designed to support higher milk production than many dairy farmers are currently achieving, most dairies with lower milk production choose to feed as much protein as was used in this example (e.g., up to 17.5 per cent crude protein of total diet dry matter for their high-producing group). Nutrient management plans usually should be based on current cow excretion estimates but, if expansion is planned, it may be helpful to estimate future needs as well.

To start, use current average production. If the herd averages 50 lbs of milk per day for all milking cows year-round (with dry cows managed on a separate farm), or if the herd averages 50 lbs year-round for all cows milking and dry (with dry cows managed in the same manure management system), then use excretion estimates for cows producing 50 lbs milk. Then multiply the excretion estimates by the average excretion per day for cows currently producing 50 lbs of milk per day by 365.

Significant volatilization of the N will occur because it is easily converted to ammonia and lost to the air as gaseous ammonia. Thus, fertilizer values will be somewhat less than amounts originally excreted. Although the value of N, P, and K fertilizer nutrients in manure usually will not be as great as the total costs of the waste management system, recovered value reduces the net cost of waste handling. However, value will be realised only if the nutrients in dairy manure are used to displace the purchase of inorganic fertilizer nutrients.

P—1.9 lbs actual P/ton wet manure (equivalent to 4.4 lbs P_2O_5)

K— 6.0 lbs actual K/ton wet manure (equivalent to 7.2 lbs K_2O)

Total solids 11.6 per cent

Manure composition changes as time passes after excretion. In addition to expected volatilization of N, liquids will drain from the solids, drying may occur, or the manure may be diluted with flushwater. Thus, the moisture content usually is quite different. It is important to take samples of manure or wastewater applied to cropland and have these samples analysed at a commercial laboratory. The analysis should include total Kjeldahl N and not just nitrate N, since almost no nitrate form of N occurs in manures. Nitrification does not occur until after manure is incorporated into the soil. The major forms of N in dairy manure are either organic N or urea N which is easily converted to ammonia and can be lost to the air as gaseous ammonia.

Manure Management System

After excretion by the cow, manure may be stored wet, stored after being allowed to dry, flushed with water to a lagoon or holding pond, spread fresh on land, or spread in some form at some later time. The longer the time in storage, the greater the potential for N losses to the air as ammonia.

The greater the dilution with water, the greater the potential for nutrient losses to surface and groundwaters unless included as part of an irrigation programme to distribute water and nutrients to growing crops. Few manure systems on farms actually collect all of the feces and urine at one location for application to one particular unit of land. Separations or losses occur in many ways.

- Flushed manure from the milking parlor and feed barn may go through a sand trap and be pumped over a separator screen before irrigation of land with the effluent.
- Manure is dropped in different areas such as pasture, milking parlor, cooling barns, and the primary feeding area and some of these "separations" may not be collectible for land-spreading.
- Some gaseous loss of ammonia occurs (volatilization) which returns a variable, but often controllable, portion of the N to the air.

Other possibilities include some surface runoff and loss to the groundwater. Management practices must control all of these components so that surface runoff and losses of nutrients to the groundwater are minimized and do not cause violations of state water quality standards.

The choice of a manure management system will depend on existing facilities. If the existing buildings were designed for flushing, then a dry handling system would not be possible without major structural modifications. If a new dairy is being planned, then other factors can be considered. In both cases, changes in the system must be compatible with other management practices on the dairy. The manure nutrients must be spread in a way that recovers nutrients in harvested crops or be stockpiled in a way that will not pose environmental risks before being spread.

Regulatory requirements may influence the choice of a waste management system. If surface runoff must be collected, stored, and dispersed on cropland, then a liquid handling system would be necessary. However, other components could still be handled as a solid or as slurry.

Types of Manure Handling Systems

Manure management systems can be categorized in many ways, with options within each type of system. The types of systems can be outlined as follows:

- Solid or conventional manure handling
- Slurry manure handling
- Liquid manure handling
- Anaerobic lagoon
- Removal of suspended solids
- Composting
- Combinations of the above

Each system can be broken down into five major components:

- Collection
- Storage
- Processing or treatment
- Transport
- Utilization

SOLID MANURE

Conventional or solid manure handling systems are not common in Florida. Normally high rainfall and humidity make it difficult to keep manure dry enough to be handled with a front-end loader. Also, the warmer climate permits the use of water for flushing on a year-round basis, as well as the utilization of wastewater on cropland in multiple cropping systems. Use of water for flushing of manure also has low labour requirements and is a clean way to handle manure.

Solid manure handling must employ some type of dry scraping for collection, and every effort must be made to keep out excess water. The manure can be hauled and spread on cropland on a daily basis or the manure can be placed in storage. The storage facility will require a roof or a provision must be made for capturing any liquid leachate or runoff in a pit or lagoon. Normally there would not be any processing or treatment of the manure. Transportation would probably be in a conventional box manure spreader, and utilization would be on cropland.

In Florida, a dairy farm would rarely attempt to employ a solid manure handling system for all of the manure. However, much of the manure can be handled as a solid. In the design of new or modified facilities, it is recommended that provisions be made for both scraping and/or flushing, if possible.

SLURRY MANURE

This is basically the mixture of excreted feces and urine with only enough water added to facilitate handling. This results in a solids content of greater than 5 per cent which is too wet to handle with a front-end loader but not dilute enough to handle in a conventional irrigation system. Storage could be in pits or in tanks either above or below ground. Transport would probably be in a tank truck or wagon or in a flail spreader. Utilization would be on cropland. If knifing or soil injection of manure nutrients is desired, this is the system of choice.

Use of slurry requires purchase of a tank truck or wagon, special pumps that will handle the high solids content, and construction of storage tanks that may be quite expensive. Loading, transporting, and spreading the slurry also has a high labour requirement but the system can result in low nutrient losses.

LIQUID MANURE HANDLING

This usually involves flush tanks to move manure and thus dilutes manure to a solids content of less than 5 per cent. In most Florida dairy systems, the solids content is less than 2 per cent and can be handled in specially designed irrigation equipment. Required storage capacity depends on type of soils in the irrigation field, the storage time between irrigation applications, and the amount of stormwater runoff which the system must be capable of retaining. Some distinct advantages of this type of system are that it has low labour requirements and can result in relatively few nutrient losses when irrigation is frequent and growing crops are available to utilize the nutrients.

ANAEROBIC LAGOON SYSTEM

This system was the most popular type of manure handling system installed on Florida dairy farms during the 1970s and '80s. It actually is a specific type of liquid handling in which the flushed manure is directed into an anaerobic lagoon. The first stage of the lagoon system is designed with a constant liquid level to overflow into a second lagoon or pond designed for liquid storage capacity.

Effluent from the storage pond must be dispersed on cropland through some type of irrigation system. Although seepage irrigation and sheet flows have been used for effluent dispersal, pivot irrigation systems are becoming much more popular because of their even distribution of the manure nutrients. In addition to the effluent, sludge accumulates in the anaerobic lagoon which contains a significant amount of P and a small amount of the N. The amount of time before the lagoon fills with sludge to a point where it needs to be cleaned out varies depending on loading rates and design volume.

Cleanout of smaller, more heavily loaded lagoons every 2 to 5 years is common while larger lagoons may function much longer before cleanout is necessary. Separation of larger manure solids is often done prior to the wastewater entering the lagoon in order to permit accommodation of manure from more cows or to extend the time until cleanout is necessary.

Removal of Suspended Solids from Flushed Manure

Moving manure from animal pens with flushed water is an easy and clean way to handle manure Variations of this method have been adopted across much of the United States in dairy, poultry, and swine operations. Separation of solids from flushed manure by some manner is potentially important in most of these systems for several reasons:

- To remove large particles and sand that would plug or damage distribution nozzles in irrigation systems used to evenly spread the liquid over the cropping area.
- (null)

To reduce the biological loading on aerobic and anaerobic lagoons.

- To capture a fibrous product and some of the N and mineral nutrients which might be utilized in other products such as bedding for free stalls, part of the feed for cattle on maintenance diets, compost for potting material for plants, etc.

Many systems exist which will remove a portion of the solids from manure slurries. The stationary screen is most common with many new dairies experimenting with settling basins of various shapes, depths, and capacities. Stationary screen separators take out 20 per cent to 30 per cent of the organic matter from flushed dairy manure. With very dilute flushed dairy manure, an estimate of 20 per cent removal of organic solids is probably most appropriate. Dilution also assures that almost all of the soluble nutrients will stay with the water portion. Most of the minerals and N are in soluble form. The feeding value of this product will not support acceptable daily gains in growing ruminants. However, the manure solids could be fed as an appreciable percentage of diets for cattle which need only to maintain themselves and sustain a slow rate of gain; well-conditioned dry cows.

Screened manure solids have been used extensively for bedding in free stalls. However, management to prepare the product properly is critical. An accepted practice seems to be to compost the solids so that internal temperatures within the pile become high enough to kill coliform bacteria. Research has shown that even though bacteria decline to low or undetectable numbers during the composting period, bacteria often return in the bedding material in the free stalls unless there is opportunity to dry the solids in sunlight.

Even when researchers found higher bacterial counts in composted dairy waste solids bedding than on rubber mats, there was no difference in bacterial counts on teats or in the milk of cows using the two types of bedding. They concluded that with "adequate" composting, dairy waste solids were a suitable bedding in free stalls. Many dairy farmers with excellent mastitis control programmes are using dry, screened manure solids for bedding in free stalls. An alternative to removing solids from flushed manure with screening equipment is to design holding basins for gravity separation (settling basins). More solids can be removed with well-designed sedimentation basins than with stationary screens. The key is the detention time of the water carrying the solids. However, the sedimented solids have much higher moisture content and are not as useful as screened solids if bedding for free stalls or composting is desired. Thus, land-spreading of these solids is the most likely method of disposal.

COMPOSTING

Composting systems have recently received much attention to facilitate exporting nutrients from dairy farms. A composting system is a modification of a conventional or solid manure handling system with the composting (treatment) process applied to the manure. Types of composting systems include windrow, bin, static pile, and aerated static pile. Major factors to be considered are the desired quality of the final product, space requirements, labour requirements, and availability of a diluent to mix with the raw waste. Before a composting system is initiated, careful attention must be given to a market or outlet for the composted manure. Do not assume that there will be a line of people waiting to buy the product. Composting is primarily an oxidative process which removes many odorous compounds. Also, some organic solids are converted to carbon dioxide thus reducing volume, and easily volatilized nitrogen is lost. However, the final product is stable and much less odorous than the initial manure.

ESTIMATED NUTRIENT LOSSES

Even with a "tightly" managed system, there is considerable N loss through ammonia volatilization. The amount volatilized is influenced by level of N in the manure (particularly the part originating in the urine) and by the method of application. Nitrogen in urine is originally excreted in the form of urea. Urease enzyme of bacterial origin is present almost everywhere manure is voided or stored so that N in urea is readily converted to ammonia which will be lost to the air as free ammonia unless the conditions of storage are acidic. It was estimated that 41 to 50 per cent of the manure N from lactating dairy cows is in urea or ammonia form (mostly from the urine).

This portion is potentially volatilized very rapidly. Most of the fecal N from cattle is in a more stable form; however, even organic N is freed and volatilized during anaerobic digestion. For irrigated, highly diluted manure the loss of ammonia during irrigation is often proportional to the evaporation loss of water. Leaching losses also may occur. Application of manures outside the growing season or in amounts which exceed crop needs may result in nitrate leaching losses of 25 per cent or more of the applied N. A high utilization of N by crops can be achieved with lowered environmental risks when manures are applied at a time when crops can absorb the mineral-N and at rates that do not exceed crop needs. Denitrification is a bacterial process which converts nitrate in solution to nitrogen gas. It is dependent upon a bacterial energy source, usually in the form of soluble organic matter, and progresses most rapidly under high moisture and/or low oxygen soil conditions. While collection and storage losses will be accounted for by sampling prior to land application, volatilization and denitrification losses after the manure leaves collection/storage are more difficult to quantify. Denitrification losses are harder to estimate on the farm but can be extensive.

POTENTIAL NUTRIENT REMOVAL BY PLANTS

One generally accepted philosophy of land application of manures is that nutrients can be applied to land in amounts slightly above the level of the nutrients removed by the crops harvested. When animal numbers are high in relation to the amount of land readily available, we need to know the maximum application rates for given soil types and crops that can be efficiently utilized by different cropping systems. We also need to use cropping systems that will take up the maximum amount of environmentally sensitive nutrients such as N and P. A long-term research project at Tifton, Georgia was designed to identify a maximum, environmentally safe application rate of manure nutrients when a triple-cropping system was used. Flushed dairy manure nutrients were applied through centre-pivot irrigation. The cropping system included Tifton 44 bermudagrass in which corn was sod-planted for silage in spring and abruzzi rye was sod-seeded in fall.

In the Georgia experiment, large-particle manure solids were separated from the liquid with an inclined stainless steel separating screen to facilitate irrigation of the effluent. The liquid portion was applied to the cropping area at four different rates. Harvests of all crops yielded 6.2 tons or more of dry matter per acre (12,400 lbs) with N-deficient application of 214 lb N/acre with yields plateauing at 12.5 to 13.0 tons of DM/acre with manure wastewater applications of 440 lb N/acre or more. Crop removals of N, however, continued to increase after DM yields plateaued because of luxury consumption of N, which increased CP and N concentrations of crops harvested. Other forage crops, even legumes, like alfalfa and perennial peanut, have been proposed as being good crops for consuming large quantities of manure nutrients. When free N is available in the soil to "scavenge," legumes take up soil N in preference to fixing N from the air. Giant elephantgrass, which has been used in field studies in Okeechobee County, Florida, gives the highest estimated P uptake. Although there may be potential for greater recovery of P in the enormous quantity of biomass harvested in giant elephantgrass than from other crops, the estimated digestible energy value of the harvested forage would be low.

The Georgia cropping system has great potential for Southern United States because: a large part of the harvest is corn silage, a high-energy forage that fits the feed needs of high producing cows; the bermudagrass forms a sod base and forage for harvest in the warm season; and rye utilizes a large amount of N during the winter season. A cropping system tested at the University of Florida that shows promise is based on perennial peanut, a higher quality base forage than bermudagrass. Perennial peanut, although not as high yielding as bermudagrass, will utilize applied N and, additionally, may offer potential as a legume to fix some N if temporary soil deficiencies occur.

The Florida research evaluated a sod-based triple-cropping system and perennial peanut overseeded with rye in winter. These were compared with a triple-crop system of corn, forage sorghum, and rye. Note that total dry matter harvests from the corn-perennial peanut-rye system were similar to those obtained with corn-bermudagrass-rye in the Georgia experiments but somewhat less N was removed, primarily due to lesser N in the winter rye crop. This was because of less dry matter yield and much less N content in the dry matter. Phosphorus removals on this system were 56 lb P/acre on all three N applications. The perennial peanut-rye system removed less dry matter but was an efficient scavenger of N, suggesting that some N-fixation from the legume may have taken place. One of the major strengths of using flushed manure systems, along with irrigation, is that additional water can be applied along with fertilizer nutrients so that full response to added nutrients is possible.

10

Pasture Use in Organic Milk Production

The most common technology used on conventional dairy operations confines milk cows in large barns and limits access to pasture. This means that forage from grazing comprises very little of the forage fed to dairy cows on most conventional dairies. Access to pasture is a requirement for the organic certification of dairy operations. Forage from grazing is also an important element of the feeding programme on many organic dairies.

Milk producers may find it easier and less expensive to maintain organic pastures for grazing dairy cows than to either purchase or produce and harvest organic crops and forage for dairy feed. Also, grazing systems may be less stressful for dairy cattle and contribute to lower veterinary expenses. Pasture-based production may also have environmental benefits, such as improved soil quality and reduced soil erosion. Pasture use for dairy feed on organic dairies, relative to conventional dairies.

More than 60 per cent of organic milk producers reported that at least half of their total forage ration came from pasture during the grazing months (an average of 6.5 months per year), compared with 18 per cent of conventional dairies. Nearly 90 per cent of organic dairies sourced at least 25 per cent of their total forage ration from pasture. In contrast, 70 per cent of conventional dairies reported that less than 25 per cent of their total forage ration fed to dairy cows came from pasture, while 40 per cent obtained none from pasture.

Notes:

> More than 60 per cent of organic dairies obtained at least 50 per cent of the forage fed to dairy cows from pasture, while nearly 70 per cent of conventional dairies obtained less than 25 per cent

To evaluate the role of pasture grazing in organic milk production, surveyed producers were divided into groups that indicate the level of pasture grazing as a source of dairy cow forage during the grazing months:

- 0-24 per cent;
- 25-49 per cent;

- 50-74 per cent; and
- 75-100 per cent.

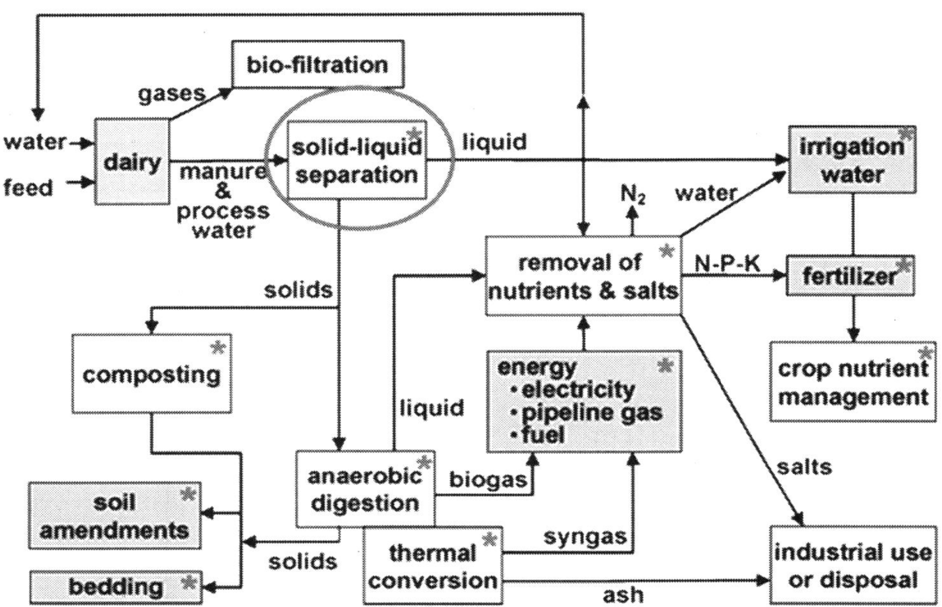

Fig. Producer Type, by Share of Dairies Forage Fed from Pasture

CHARACTERISTICS AND PRACTICES BY LEVEL OF PASTURE USE

Almost two-thirds of organic milk producers reported that at least 50 per cent of the forage fed to dairy cattle during the grazing months came from pasture, and a third reported that 75 per cent or more came from pasture. Only 12 per cent of organic milk producers reported that less than 25 per cent of forage fed came from pasture.

Differences between organic dairies where pasture use was highest (75-100 per cent of forage fed) and lowest (0-24 per cent of forage fed) were significant and are emphasized in this section. Farm characteristics among organic dairies in the middle groups (25-49 per cent and 50-74 per cent of forage fed from pasture) showed little difference. Operations that relied on pasture use the most were the smallest in size and produced the least milk per cow. Organic dairies with 75-100 per cent of forage fed from pasture had an average of 64 milk cows that averaged 11,289 pounds of milk per cow.

In contrast, dairies with 0-24 per cent of forage from pasture averaged more than twice the number of milk cows (135) and more than 5,000 pounds more milk production per cow (16,560 pounds). Thus, operations that relied less on pasture use (12 per cent) had a disproportionately large share of milk cows (21 per cent) and milk production (25 per cent), while a third of operations that relied most on pasture use accounted for only 25 per cent of milk cows and 21 per cent of production.

Table. Characteristics and Production Practices, by Level of Pasture use on Organic Dairy Operations, 2005

	Forage Fed from Pasture During Grazing Months			
Item	**0-24** %	**25-49** %	**50-74** %	**75-100** %
Per cent of farms	12	24	32	31
Per cent of milk cows	21	23	31	25
Per cent of milk production	25	23	31	21
Milk cows *(number per farm)*	135	79	78	64
Milk production *(pounds per cow)*	16,560	13,594	13,319	11,289
Region:	*Per cent of farms*			
Northeast (ME, NY, PA, VT)	21	42	43	55
Upper Midwest (MI, MN, WI)	58	44	40	35
Corn Belt (IL, IN, IA, MO, OH)	13	8	5	9
West (CA, ID, OR, WA)	8	6	12	2
Region:	*Per cent of cows*			
Northeast (ME, NY, PA, VT)	9	29	28	45
Upper Midwest (MI, MN, WI)	30	46	28	29
Corn Belt (IL, IN, IA, MO, OH)	10	9	4	8
West (CA, ID, OR, WA)	51	16	40	18
In dairy business:	*Per cent of farms*			
Less than 10 years	11	18	19	27
10 years or more	89	82	81	73
Pasture rotation:				
At least once a day	39	48	58	72
Every 2-7 days	30	33	38	26
Less than weekly	4	1	id	1
Never	24	18	3	1
Pasture not used for feed	4	0	0	0

Notes:

id=Insufficient data for disclosure.

Organic dairies with the highest pasture use for feed were located in the Northeast. More than half of these farms were in the Northeast and operated with 45 per cent of the organic milk cows in this group.

Operations in the Upper Midwest accounted for the majority of farms using the least pasture (0-24 per cent) and 35 per cent of farms using the most (75-100 per cent). Organic dairies in the West accounted for only 8 per cent of the farms using the least pasture, but these large operations held 51 per cent of the milk cows in this group. Large operations relied less on pasture as a forage source because of the significant land requirements necessary to supply pasture for large herds. However, some large operations in the West used pasture as an important dairy feed source. Among the farms with the highest pasture use, the West (at 2 per cent) included 18 per cent of the milk cows. Organic dairies that relied the most on pasture use were in business relatively less time than other organic dairies.

Twenty-seven per cent of dairies that relied on pasture for 75-100 per cent of their forage feed had been in business less than 10 years, compared with only 11 per cent that relied on pasture for 0-24 per cent of their forage feed. Also, farm operators who relied the most on pasture forage were younger than other farm operators. Sixteen per cent of these farm operators were younger than 35 and only 4 per cent were older than 65, compared with 6 and 10 per cent, respectively, of the operators who used the least pasture. Younger farm operators on relatively new dairies were more likely to use pasture-based feeding for organic milk production.

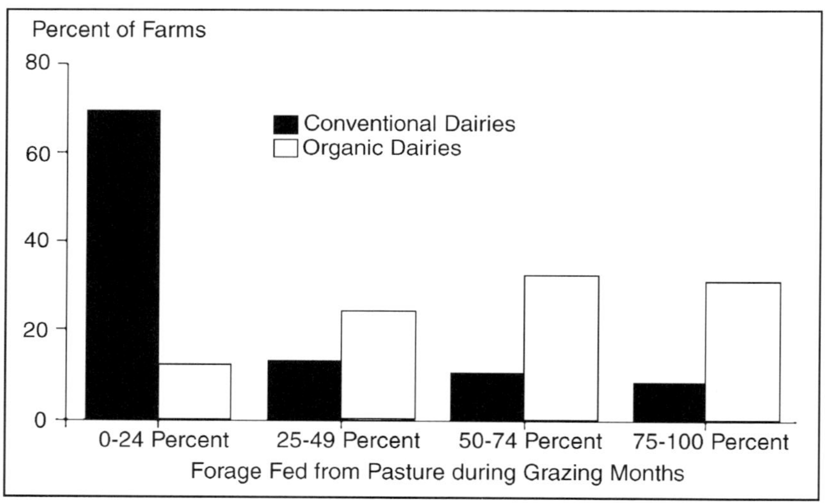

Fig. Farm Operator Age, by the Share of Dairy Forage Fed from Pasture

Notes:

Among organic farm operators that used the most pasture for dairy feed, 16 per cent were younger than 35 and only 4 per cent were older than 65

Pasture as a source of forage on organic dairies was closely related to the intensity of pasture management. Rotating pastures is a management strategy that can increase the total volume and quality of dry matter produced on pastures by allowing cows to graze for specified periods before moving them to new pastures. Seventy-two per cent of organic dairies that reported the highest use of pasture forage rotated pastures at least once a day, compared with less than 40 per cent of dairies that used pasture forage the least. Among dairies that used pasture the least, 24 per cent reported that pastures were never rotated and 4 per cent indicated that pasture was not a source of dairy feed.

PRODUCTION COSTS AND NET RETURNS BY LEVEL OF PASTURE USE

Substituting pasture for more expensive feed sources may appear to be an efficient way to lower dairy feed costs. Average feed costs per cwt of milk, however, were lowest on the organic operations using the least pasture for dairy forage (0-24 per cent), between $1 and $2 per cwt less than on the operations using more pasture.

Table. Production Costs and Net Returns on Organic Dairy Operations, by Level of Pasture use, 2005

	Forage Fed from Pasture During Grazing Months			
Item	**0-24** %	**25-49** %	**50-74** %	**75-100** %
	Dollars per cwt sold			
Feed:	11.61	13.14	13.42	12.90
Purchased feed	7.43	5.15	7.19	6.09
Homegrown harvested feed	3.89	7.51	5.41	5.69
Grazed feed	0.29	0.48	0.82	1.13
Hired labour	2.44	1.85	2.17	2.27
Capital costs	4.11	5.36	4.48	6.70
Opportunity cost of unpaid labour	2.82	6.47	6.24	8.36
Cost summary				
Operating costs	17.74	18.55	18.99	19.44
Operating and capital costs	21.85	23.91	23.47	26.14
Total economic costs	25.33	31.53	30.64	35.99
Gross value of production	24.10	24.49	23.59	25.63
Returns above total economic costs	-1.22	-7.04	-7.05	-10.36

Notes:

Measuring Milk Production Costs. The gross value of production includes milk sales, cull and breeding animal sales, revenue from leasing dairy animals or space to other operations, co-op patronage dividends, dairy assessment rebates or refunds, and the value of manure. The value of milk sales was determined from prices reported by the survey respondents.

This relationship changes little even if charges for pasture resources are excluded from the feed costs. Average feed costs per cow declined as pasture use for dairy forage increased.

Total feed costs were $1,902 per cow on organic dairies that relied on pasture for 0-24 per cent of forage fed, compared with only $1,409 on organic dairies that relied on pasture for 75-100 per cent of forage fed, a savings of about 25 per cent. As shown previously, however, average production per cow was more than 30 per cent lower on the organic dairies that used the most pasture, thus total costs per cwt were higher.

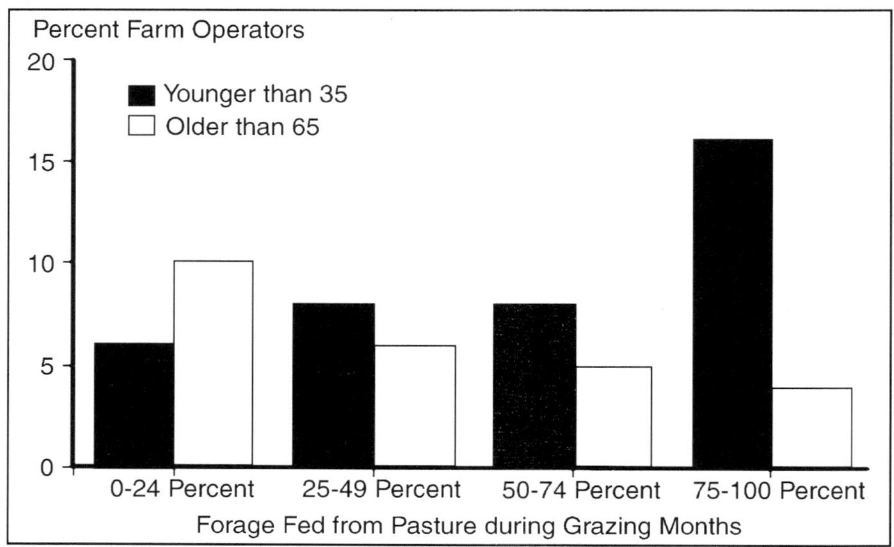

Fig. Feed Costs, by Share of Dairy Forage Fed from Pasture

Notes:

Average feed cost per cow on organic dairies were about 25 per cent less on operations using the most pasture.

Breaking down the distribution of milk production per cow for each group provides an indication of how milk yields declined as pasture was substituted for higher energy feed in the dairy ration. Nearly half of organic dairies feeding 0-24 per cent of forage from pasture had annual milk yields at 15,000 pounds per cow or more, compared with just 10 per cent of dairies feeding 75-100 per cent.\

In contrast, more than 40 per cent of dairies using the most pasture had an annual milk yield of less than 10,000 pounds per cow, compared with 19 per cent of those using the least pasture. This suggests that improving pasture quality to achieve higher milk production can contribute to the success of organic dairies using pasture-based feeding. The relationship between pasture use and total production costs follows much the same pattern as seen for size and production costs on organic dairies. That is, the operations using less pasture were much larger and had lower average costs per cwt than the smaller operations using more pasture.

Capital costs were more than $2 per cwt higher and the charge for unpaid labour was more than $5 per cwt higher on the operations where pasture forage accounted for 75-100 per cent of forage fed than on those where pasture forage accounted for 0-24 per cent of forage fed. Larger operations were able to spread the fixed costs for capital and labour over production from more cows, and each cow was more productive on the larger operations using the least pasture for dairy forage.

Average returns net of total economic costs were negative for dairies in each of the pasture use groups, but were significantly higher among the dairies using the least pasture. This disparity suggests that organic operations using conventional dairy feeding methods, such as confining cows and feeding higher energy feed, were more likely to generate higher returns to capital and labour resources than those relying more on pasture-based feeding.

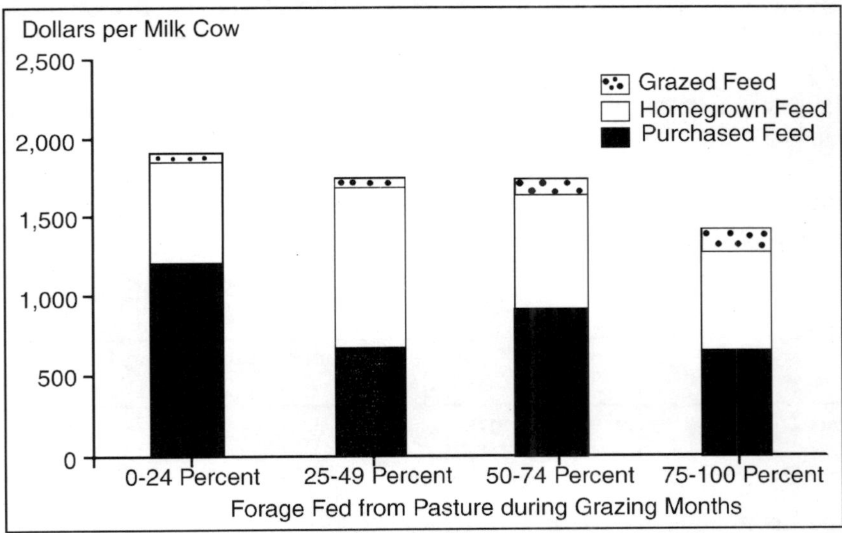

Fig. Milk Production, by Share of Dairy Forage Fed from Pasture

Notes:

Nearly half of organic dairies feeding the least pasture forage had milk yields above 15,000 pounds per cow, compared with only 10 per cent of those feeding the most pasture.

Thus, the technologies and production methods used on organic dairies could become more like those used on conventional dairies. Changing technologies and production methods may depend on the extent to which pasture requirements for organic dairies must include pasture as part of the dairy feeding programme.

LABOUR USE BY LEVEL OF PASTURE USE

The biggest difference in organic dairy costs between those using the most and the least pasture for dairy forage was the unpaid labour charge. The labour charge on organic dairies feeding 75-100 per cent of forage from pasture was more than $8 per cwt, compared with less than $3 per cwt on dairies feeding 0-24 per cent of forage from pasture. The labour charge contributed about half the difference in total costs that were more than $10 per cwt higher for dairies using the most pasture. Table includes labour use per cow and per cwt for each level of pasture feeding.

Unpaid labour use per cwt on organic dairies feeding 75-100 per cent of forage from pasture was about three times that for those feeding 0-24 per cent of forage from pasture. Part of this difference came from fixed amounts of labour being spread over more output on the larger operations that used the least amount of pasture. Some of the labour difference, however, may be due to the pasture feeding system's being more labour intensive than handling dairy cows in confinement.

Among farms with fewer than 100 cows, the unpaid labour use on dairies feeding the least forage from pasture was still about a third less than on those feeding the most forage from pasture (0.45 hour versus 0.65 hour). In addition, total labour use among these small dairies increased steadily as more pasture was used for dairy feed. This increase suggests that pasture-based feeding systems are inherently more labour intensive than the conventional practice of handling dairy cows in confinement, possibly due to the labour required for moving cows to, from, and between pastures.

Table. Organic Dairy Labour Use, by Level of Pasture Use, 2005

	Forage Fed from Pasture During Grazing Months			
Item	**0-24** %	**25-49** %	**50-74** %	**75-100** %
Per cow				
Hired labour hours	30	20	22	17
Unpaid labour hours	27	52	48	56
Total labour hours	57	72	70	73
Per cwt				
Hired labour hours	0.18	0.15	0.17	0.15
Unpaid labour hours	0.16	0.39	0.36	0.49
Total labour hours	0.34	0.53	0.52	0.65
Per cwt—dairies with fewer than 100 cows				
Hired labour hours	0.15	0.18	0.17	0.16
Unpaid labour hours	0.45	0.52	0.59	0.65
Total labour hours	0.60	0.70	0.76	0.81

Note:

Totals may not equal the sum of items due to rounding

ECONOMIES OF SIZE IN ORGANIC MILK PRODUCTION

The emergence of large conventional dairy farms and the continued shift of production towards such farms suggest that economies of size play a major role in the structure of dairy farming. Economies of size and the role it takes in the structural changes observed in conventional milk production were documented by MacDonald *et al.* This section reports on the relationship between operation size and production costs in organic milk production, including what size and costs might mean for the structure of the industry.

To evaluate the relationship between organic milk production costs and operation size, surveyed producers were divided into groups by size, and differences in farm characteristics and practices, production costs and net returns, and labour use were compared among the groups.

The size groups were defined by the largest number of cows milked on the operation during 2005:

- Fewer than 50 cows;
- 50-99 cows;
- 100-199 cows; and
- 200 cows or more.

CHARACTERISTICS AND PRACTICES BY SIZE

Data for 2005 indicated that about 45 per cent of organic dairies milked fewer than 50 cows, while 87 per cent had fewer than 100 cows. These small operations accounted for a disproportionately small share of organic milk cows and production.

In contrast, about 13 per cent of organic dairies had 100 cows or more, but they accounted for 44 per cent of milk cows and nearly half of production. The largest organic dairies—those with 200 cows or more—had an average herd size of almost 500 cows per farm and included nearly a third of organic milk cows. These large dairies had a higher average milk yield per cow than smaller dairies and produced 37 per cent of organic milk.

The most striking difference among organic dairies in each size group was their location. Northeast and Upper Midwest farms accounted for 96 per cent of dairies with fewer than 50 cows. Farms in the West accounted for 80 per cent of dairies with 200 cows or more.

Within each region, 97 per cent of organic dairies in the Northeast, 88 per cent of organic dairies in the Upper Midwest, and 85 per cent of organic dairies in the Corn Belt had fewer than 100 cows. In contrast, 83 per cent of organic dairies in the West had 100 cows or more and 63 per cent had 200 cows or more.

The use of several milk production technologies and practices tended to increase with size of organic operation, but most of the differences were between the smallest and other organic farms. For example, use of a milking parlor increased from 22 per cent for the smallest dairies to nearly all dairies with 100 cows or more. Thirty-eight per cent of the smallest farms and over 50 per cent of other farms participated in the Dairy Herd Improvement programme.

Table. Dairy Farm Characteristics and Production Practices, by Size of Organic Operation, 2005

	Size of Organic Dairy				
Item	**Fewer than 50 Cows**	**50-99 Cows**	**100-199 Cows**	**200 Cows or More**	
% of farms	45	42	8	5	
% of milk cows	21	35	13	31	
% of milk production	18	33	12	37	
Milk cows (number per farm)	37	68	132	490	
Milk production (pounds per cow)	11,884	12,796	12,008	16,133	
% of farms					
Region:					
Northeast (ME, NY, PA, VT)	51	46	17	0	
Upper Midwest (MI, MN, WI)	45	39	52	20	
Corn Belt (IL, IN, IA, MO, OH)	4	12	15	0	
West (CA, ID, OR, WA)	0	3	16	80	
Production practices:					
DHI programme participation[1]	38	51	53	60	
Pasture-based feeding[2]	70	60	49	54	Artificial
insemination	70	76	71	87	
Embryo transplants or sexed semen	3	3	6	2	
Controlled breeding/calving season	38	35	36	20	
Regular veterinary services	28	41	72	55	
Nutritionist services	30	53	66	70	
Computerized milking system	1	4	4	7	Computerized
feeding system	id	3	5	31	
Milking parlor	22	45	88	100	
Kept individual cow records	59	64	62	70	
Johne's disease programme participation	18	28	44	45	
Onfarm computer records	13	25	34	41	
Dairy information from Internet	30	46	56	64	
Forward-purchased inputs	4	10	22	16	
Negotiate input price discounts	18	21	33	36	

Notes:

id=Insufficient data for disclosure.

[1]Dairy Herd Improvement.

[2]Pasture-based feeding is defined as providing at least half of the forage fed to milk cows during the grazing months from pasture. Organic dairies reported an average grazing period for milk cows of 6. 5 months in 2005.

Use of regular veterinary and nutritionist services was also much lower on the smallest farms. Likewise, the use of business management tools, such as computer records, the Internet, and forward-purchasing inputs, was less. Pasture-based feeding was used by 70 per cent of the smallest dairies and was also used by more than half of the largest dairies.

PRODUCTION COSTS AND NET RETURNS BY SIZE

Average feed costs differed little among the dairies in each size group, but average operating costs for organic dairies increased from the smallest to largest size group due to the additional costs of hired labour on the largest farms.

Total operating costs were about $2 per cwt higher on the largest organic dairies than on the smallest. Once capital costs were added, cost differences among the size groups were negligible, as lower capital costs on the larger operations offset the additional hired labour costs.

The smallest farms, however, used significantly more unpaid labour, mainly from farm operators and other family members, than did larger farms. After adding an opportunity cost for this unpaid labour, total economic costs for organic dairies declined with size and were nearly $14 per cwt less on the largest farms than on the smallest.

Table. Production Costs and Returns, by Size of Organic Dairy Operation, 2005

	Size of Organic Dairy			
Item	**Fewer than 50 Cows**	**50-99 Cows**	**100-199 Cows**	**200 Cows or More**
Dollars per cwt sold				
Feed:	12.96	12.67	11.93	13.06
Purchased feed	5.16	4.94	5.29	8.99
Homegrown harvested feed	6.96	7.00	6.06	3.51
Grazed feed	0.83	0.73	0.58	0.56
Hired labour	0.79	1.75	2.25	3.21
Capital costs	6.02	6.20	5.04	3.58
Opportunity cost of unpaid labour	13.42	7.45	4.84	1.15
Cost summary:				
Operating costs	17.65	18.25	18.10	19.66
Operating and capital costs	23.67	24.44	23.13	23.25
Total economic costs	38.50	33.36	29.13	24.84
Gross value of production	25.57	24.88	23.46	23.59
Returns above total economic costs	-12.93	-8.48	-5.67	-1.26

Notes:

Costs are defined in Appendix A: Measuring Milk Production Costs. The gross value of production includes milk sales, cull and breeding animal sales, revenue from leasing dairy animals or space to other operations, co-op patronage dividends, dairy assessment rebates or refunds, and the value of manure. The value of milk sales was determined from prices reported by the survey respondents.

The relationship between average production costs and operation size for organic dairies was similar to that for conventional dairies among the size groups that characterize the organic industry.

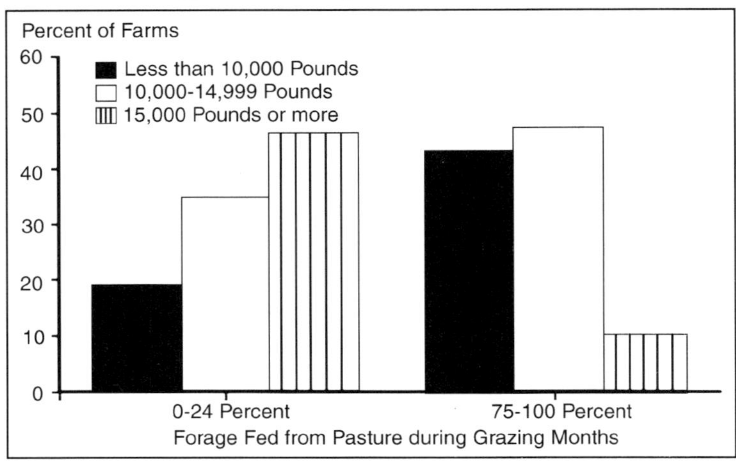

Fig. Total Economic Costs for each Type of Dairy Producers, by Farm Size

Notes:

The relationship between production costs and operation size for organic dairies was similar to conventional diaries, only at higher costs

Costs shown for conventional dairies of 200 cows or more are reported for 200-499 cows by MacDonald *et al.*, with an average size of 295 cows. The average size of organic dairies in the 200 cows or more group is 490 cows.

Average production costs for conventional milk production declined from about $30 per cwt on operations with fewer than 50 cows to about $18 per cwt on operations with 200-499 cows. Average costs dropped to less than $14 per cwt on conventional operations with more than 1,000 cows. Lower costs among the largest conventional producers suggest that similar economies of size may be available to organic producers if they increase the size of their operation above the industry norm.

However, the additional costs of complying with pasture requirements and securing organic inputs in large volume may limit the cost advantages for larger organic operations. Net returns to the organic dairy enterprise were computed as the difference between the gross value of production and total economic costs. Average net returns in 2005 were negative for all organic dairy size groups (from-$12.93 per cwt among dairies with fewer than 50 cows to-$1.26 per cwt among dairies with 200 cows or more). These averages, however, mask the variation in net returns among dairies. Some operations may be more productive because they are well managed, while others may pay less than average for inputs or receive above-average milk prices, and thus these farms may be profitable while their group, on average, is not. For example, 10 per cent of organic dairies with fewer than 50 cows and almost half of those with 200 cows or more generated positive returns above total economic costs.

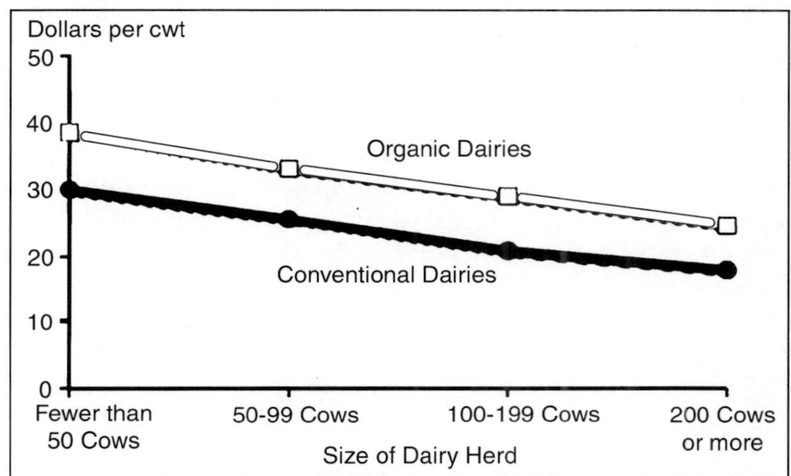

Fig. Organic Dairy Operations with Positive Returns, by Size of Operation

Notes:

Ten Per cent of the smallest and 47 per cent of the largest organic dairies have positive returns above total costs, but many more have positive returns when capital costs are excluded.

Farmers are unlikely to start or expand a dairy operation if they are unable to recover their investment in capital and labour. Farmers who have already made substantial capital investments in dairy facilities and equipment, however, with little or no opportunity for use outside the dairy operation. These costs, therefore, have no bearing on the decision to operate in the short term, only becoming important when these capital assets must be replaced.

To better measure costs relevant to shortterm operating decisions, the share of farms with gross returns exceeding all costs except for capital costs was computed. About a fourth of organic dairies with fewer than 50 cows covered these costs, compared with almost threefourths of those with 200 cows or more. Some small organic dairy farms appeared to be earning enough to keep operating, but on average, farms in the smaller size groups were not covering the opportunity costs of their investments in capital and the operator's time.

In contrast, larger organic dairies were much more likely to generate returns above capital and labour costs. These conditions suggest that organic milk production may migrate towards larger operations, as has conventional production. This production change will likely occur over an extended period as existing smaller operations use up their production facilities and equipment, ultimately facing decisions about replacing capital assets or exiting the industry. Also, in the short term, some small operators may continue producing organic milk as a lifestyle choice despite returns that fail to cover the opportunity cost of their time.

LABOUR USE BY SIZE

Small organic dairies are at a cost disadvantage relative to larger dairies due to the opportunity cost assigned to the unpaid labour provided by operators, partners, and family members. Unpaid labour costs for organic dairies with fewer than 50 cows were $13.42 per cwt, compared with only $1.15 per cwt on dairies with 200 cows or more. If this opportunity cost is ignored, the net returns to organic milk production were more similar among the size groups. The largest dairies used substantially more hired labour than the smallest (11,530 hours versus 557 hours), but the hired labour hours per unit of production differed little among the size groups because of much higher production on the largest dairies.

The difference in unpaid labour use among the farm size groups was much less. Organic dairy operations with 200 cows or more used about 700 more hours of unpaid labour, about 20 per cent more than those with fewer than 50 cows. The largest dairies, however, produced about 1,700 per cent more milk as a result of having more cows and an average productivity per cow that was 36 per cent higher than the smallest dairies. Enhanced productivity for these larger dairies means that unpaid labour hours per unit of production fell dramatically across the size groups from 0.82 hour per cwt on the smallest organic dairies to only 0.05 hour on the largest.

Table. Labour Use, by Size of Organic Dairy Operation, 2005

	Size of Organic Dairy			
Item	**Fewer than 50 Cows**	**50-99 Cows**	**100-199 Cows**	**200 Cows or More**
Per farm				
Hired labour hours	557	1,648	3,051	11,530
Unpaid labour hours	3,612	3,759	4,265	4,322
Total labour hours	4,169	5,407	7,316	15,852
Milk production (cwt)	4,367	8,748	15,825	78,977
Per cwt				
Hired labour hours	0.13	0.19	0.19	0.15
Unpaid labour hours	0.82	0.43	0.27	0.05
Total labour hours	0.95	0.62	0.46	0.20

Unpaid labour is an opportunity cost defined as the time spent working on a dairy operation charged at a wage rate that represents what unpaid workers could earn in off-farm employment.

Opportunity costs of labour may vary significantly among producers, and some producers may be willing to accept returns lower than they could earn in nonfarm employment because of lifestyle preferences and costs of switching occupations, among other reasons. The incentive to earn a competitive return for the time spent on the dairy operation, however, may be a motivating force for increasingly larger organic dairies.

CURRENT DAIRY POLICIES TO ASSIST PRODUCERS

Indian dairy policy has been developed over the last seven decades.

The early policies addressed three main problems:

1. Producers lacked bargaining power with milk buyers;
2. Producers suffered from volatile or low prices; and
3. Market participants encountered severe shortages/gluts resulting from marketing a highly perishable commodity.

The policy response resulted in the development of two major government activities that still function today: federal milk marketing orders and the Dairy Product Price Support Programme.

While both FMMOs and the DPPSP have their roots in the 1930s and 1940s, the programmes have changed modestly over the years as the industry structure and markets changed. Two other components of Indian dairy policy are relatively new programmes. First, the 1985 farm bill established the Dairy Export Incentive Programme to counter foreign competitor subsidies.

Second, the Milk Income Loss Contract programme was established in the 2002 farm bill as a government payment for dairy farmers in times of low milk prices. Like Indian crop programmes, the MILC programme pays dairy producers when prices decline below a specified level. The following sections describe each of these four components and how they relate to the current market situation.

Lower milk and dairy product prices since late 2008 have generated new programme activity. Purchasing dairy products last fall under the DPPSP; MILC payments were triggered beginning in February.

MILK INCOME LOSS CONTRACT PROGRAMME

The Milk Income Loss Contract (MILC) programme pays dairy farmers when farm milk prices fall below an established target price. Section 1506 of the 2008 farm bill extends authority for the MILC programme until September 30, 2012. This programme is similar to long-time subsidy programmes for crops that pay farmers when farm prices drop below certain levels. Farm Service Agency implements the MILC programme. Under MILC, participating dairy farmers nationwide are eligible for a federal payment whenever the minimum monthly market price for farm milk used for fluid consumption in Boston falls below $16.94 per cwt.

Eligible farmers then receive a payment equal to 45 per cent of the difference between the $16.94 target price and the lower monthly market price. The payment quantity is limited to 2.985 million pounds of annual production. Since the inception of the MILC programme, large dairy farm operators have expressed concern that the payment limit has negatively affected their income. For larger farm operations, their annual production is well in excess of the limit, and any production in excess of that receives no federal payments.

To address the issue of rising feed costs, the 2008 farm bill includes a provision that adjusts upward the $16.94 target price in any month when feed prices are above a certain threshold. The law requires calculating monthly a National Average Dairy Feed Ration Cost based on a formula that currently uses to calculate feed costs. In any month that the average feed cost is above $7.35 per cwt., the $16.94 target price will be increased by 45 per cent of the difference between the monthly feed cost and $7.35. For the latter half of 2007 and all of 2008, farm milk prices remained well above the MILC trigger price, precluding the need for any MILC payments.

However, milk prices have since declined below the trigger for MILC payments. The Class I Boston farm milk price for February 2009 was $13.97 per cwt. With the adjustment for feed costs raising the trigger to $17.33 per cwt., MILC payments were activated for the first time in two years at a payment rate of $1.51. The payment rate rose to $2.01 per cwt. in March. Given current prospects in the futures markets for milk, corn, and soybeans, payments are expected to continue during 2009, but at smaller rates.

Individual producers must select which month to begin receiving payments, based on their projection of potential payment rates and the possibility of hitting the production payment limit. As of October 26, 2009, total MILC payments distributed to date were $775 million. The timing of the payments has caused some concern for producers this spring. While milk price data become available during the payment month, data needed for the feed cost adjustor are not available until publishes monthly average feed prices in *Agricultural Prices* at the end of the next month. Consequently, MILC payments for a particular month are not processed until two months later.

DAIRY PRODUCT PRICE SUPPORT PROGRAMME (DPPSP)

The Agricultural Act of 1949 first established a dairy price support programme by permanently requiring supporting the farm price of milk. Since 1949, Congress has regularly amended the programme, usually in the context of multiyear omnibus farm acts and budget reconciliation acts. Historically, the supported farm price for milk is intended to protect farmers from price declines that might force them out of business and to protect consumers from seasonal imbalances of supply and demand.

Commodity Credit Corporation (CCC) supports milk prices by its standing offer to purchase surplus non-fat dry milk, cheese, and butter from dairy processors. Whenever market prices fall to product support levels, processors generally make the business decision of selling surplus product to the government rather than to the marketplace. Consequently, the government purchase prices usually serve as a floor for the market price, which in turn indirectly supports the farm price of milk for all dairy farmers.

The effectiveness of the dairy price supports depends on removal of products from the market and placement into government storage. The Dairy Product Price Support Programme (DPPSP) as authorized by the 2008 farm bill requires purchasing products at the following minimum prices: block cheese, $1.13/lb.; barrel cheese, $1.10/lb.; butter, $1.05/lb.; and non-fat dry milk, $0.80/lb.

Under previous law, the support price for farm milk was statutorily set at $9.90 per cwt., and given the administrative authority to establish a combination of dairy product purchase prices that indirectly supported the farm price of milk at $9.90. Although the 2008 law does not specifically state that the overall support price is $9.90 per cwt, each of the mandated product prices in the law is equivalent to the existing product purchase prices, so farm milk prices effectively continue to be supported at $9.90.

In late 2008 and 2009, after several years of relative inactivity, the price support programme resumed purchases when dairy product prices approached support levels. As of September 11, 2009, estimated that it purchased 111 million pounds of non-fat dry milk under the programme in 2008 and expects to purchase 379 million pounds in 2009, along with small amounts of butter and cheese (including amounts exported under the Dairy Export Incentive Programme).Total expenditures on the DPPSP were $223 million from October 1, 2008, through September 10, 2009. With an expected rise in milk and product prices next year, Forecasts only a small amount of butter to be purchased in 2010.Following heightened industry and congressional interest in taking action to boost milk prices for farmers, July 31, 2009, a temporary increase in price support for cheese and non-fat dry milk from August 2009 through October 2009.

Subsequently, the Senate approved an amendment to the Senate-passed FY2010 agriculture appropriations bill to increase Farm Service Agency funding by $350 million, ostensibly for an additional increase in dairy product price support levels. However, the conference agreement for the FY2010 Agriculture appropriations bill, which was enacted on October 21, 2009, provides for a different use of the funds.

MILK MARKETING ORDERS

Federal milk marketing orders (FMMOs) mandate minimum prices that processors must pay producers for milk depending on its end use. This compares with the MILC programme, which provides direct payments to producers, and the DPPSP, which buys surplus dairy products at specified minimum prices. The DPPSP serves as a price floor for products and under girds FMMO minimum milk prices. The farm price of approximately two-thirds of the nation's fluid milk is regulated under FMMOs.

Federal orders, which are administered Agricultural Marketing Service, were instituted in the 1930s to promote orderly marketing conditions by, among other things, applying a uniform system of classified pricing throughout the market. Some states, California for example, have their own state milk marketing regulations instead of federal rules. FMMOs also address how market proceeds are distributed among producers delivering milk to federal marketing order areas.

Producers are affected by two fundamental marketing order provisions: the classified pricing of milk according to its end use, and the pooling of receipts to pay all farmers a blend price. Federal orders regulate dairy handlers (processors) who sell milk or milk products within a defined marketing area by requiring them to pay not less than established minimum class prices for the Grade A milk they purchase from dairy producers, depending on how the milk is used.

This classified pricing system requires handlers to pay a higher price for milk used for fluid consumption than for milk used in manufactured dairy products such as yogurt, ice cream, and sour cream, cheese and butter and dry milk products. These differences between classes reflect the different market values for the products.

Blend pricing allows all dairy farmers who ship to the market to pool their milk receipts and then be paid a single price for all milk based on order-wide usage (a weighted average of the four usage classes). Paying all farmers a single blend price is seen as an equitable way of sharing revenues for identical raw milk directed to both the higher-valued fluid market and the lowervalued manufacturing market. Manufactured class prices are the same in all orders nationwide and are calculated monthly based on current market conditions for manufactured dairy products. The Class I price for milk used for fluid consumption varies from area to area.

Class I prices are determined by adding, to a monthly base price, a "Class I differential" that generally rises with the geographical distance from milk surplus regions in the Upper Midwest, the Southwest, and the West. Class I differential pricing is a mechanism designed to ensure adequate supplies of milk for fluid use at consumption centers. The supply of milk may come from local supplies or distant supplies, whichever is more efficient. However, local dairy farmers are protected by the minimum price rule against lower-priced milk that might otherwise be hauled into their region. Over the years, dairy farmers have supported minimum prices afforded by FMMOs because they help balance marketing power traditionally held by processors. In contrast, dairy processors generally oppose them. Mandated minimum prices, they say, do not allow for timely adjustments in a rapidly changing market and can leave product manufacturers in unprofitable situations. Also, they contend that the FMMO system distorts markets, saying fixed differentials contributed to high fluid milk prices last year.

DAIRY EXPORT INCENTIVE PROGRAMME (DEIP)

First authorized in 1985, the Dairy Export Incentive Programme (DEIP) provides cash bonus payments to Indian dairy exporters. The programme was initially intended to counter foreign—mostly European Union—dairy subsidies (while removing surplus dairy products from the market), but subsequent farm bill reauthorizations have added market development to the role of DEIP. Payments since the program's inception have totaled $1.1 billion. The programme was active throughout the 1990s, peaking in 1993 with $162 million in bonuses.

DEIP funding is a mandatory account provided through the Commodity Credit Corporation (CCC) borrowing authority from the Indian Treasury, rather than through annual appropriations bills. The programme had not been used since FY2004 until announced its reactivation on May 22, 2009. Indian dairy product exports made with DEIP bonuses are subject to annual limitations under the Uruguay Round Agreement of the World Trade Organization (WTO).

The limits are 68,201 metric tons of skim milk powder, 21,097 tons of butterfat, 3,030 tons of various cheeses, and 34 tons of other dairy products (quantity limits are on a July-June year). Total expenditures under WTO commitments are now capped at $117 million per year (value limits on a October-September year).

COMPONENTS OF MILK FOOD

Milk contains components that are essential to humans such as proteins, carbohydrates, fat, water, all the B-vitamins, vitamins A and D, calcium and phosphorus. It also provides energy.

An important protein in milk is casein (in many cases 80% of the milk protein). This is the base for cheese making. Casein is linked to calcium phosphate, which is why milk contains a relatively large amount of this salt that is a very important nutrient for humans and animals. In addition to casein, milk contains whey proteins (20% of the milk protein). The whey proteins are in most cases not incorporated in the cheese; they remain in the whey. Whey proteins (globulins and albumins) have a very high nutritive value.

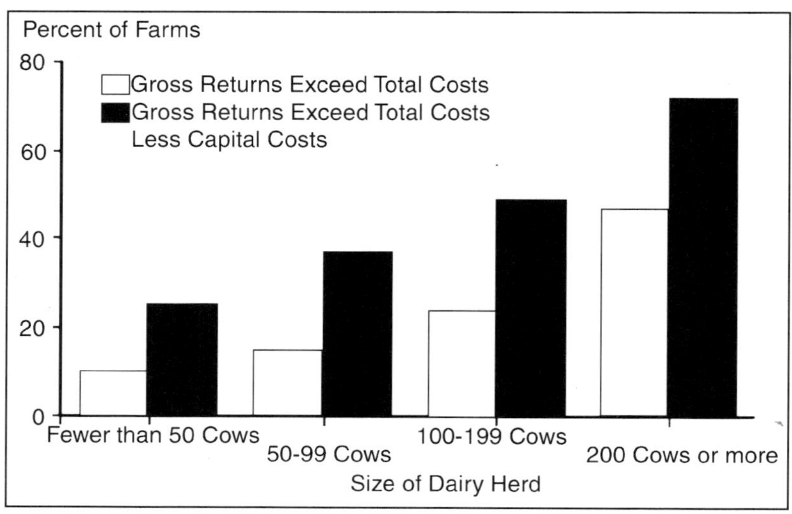

Fig. Main Components of Milk

Milk protein is of a high quality. This means that the human body can use a large part of the protein efficiently. Proteins in various other foodstuffs have a comple-mentary effect. In combination with cereals, potatoes, meat, eggs or nuts in one meal, the body can use an even greater percentage of the milk protein. Apart from milk, there are other animal protein sources such as fish and meat.

Vegetable protein, which is also important in making the body's proteins, is found in cereals and pulses. Protein is needed by the body for growth, replacement of worn-out body proteins and the production of compounds that the body needs. Milk sugar (lactose) is a carbohydrate, a necessary component to keep the body going. Our bodies burn carbohydrates in the same way an oven burns wood. Through this combustion, energy is released which is used by our bodies for many kinds of activities.

Milk fat is present in the form of small fat globules, which have a lower weight than the other components of the milk. When cow milk is allowed to stand, these globules collect on top of the milk and form a layer of cream. Buffalo milk also forms some cream on top, but other kinds of milk, such as that of sheep and goats, hardly form a layer of fat at all. For these types of milk one needs to separate the cream from the milk.

Milk fat is easy to digest. The body uses fat as a fuel or stores it as fat reserves. Milk is also an important source of minerals and vitamins. It contains large quantities of calcium, which can easily be absorbed by the body after digestion and is important for the formation of bones (the skeleton).

Milk is also an important source of vitamin B2 (Riboflavin), but there is little vitamin C in milk. Therefore a person's diet must also include vegetables and fruits in order to ensure a sufficient supply of vitamin C.

Milk is able to compensate for a lack of certain nutrients in a monotonous diet because of the great diversity of nutrients it contains and the high value of milk protein. It can therefore greatly improve the quality of the diet.

Products derived from milk contain these nutrients to a greater or lesser extent. Milk is especially desirable for vulnerable groups, for instance babies, toddlers, children and pregnant and nursing mothers. Always strive for a healthy, varied diet, which apart from milk also includes cereals, pulses, vegetables, fruits and if possible meat or fish. The various types of milk differ in various ways, including nutritional value.

Bibliography

Ajay Kapoor: *Milk and Its Products*, Vishvabharti Publication, Delhi, 2005.

Alan H. Varnam and Jane P. Sutherland: *Milk and Milk Products : Technology, Chemistry and Microbiology*, Springer Publication, Delhi, 2010.

Andrew L. Winton and Kate Barber Winton: *Milk and Milk Products*, Agrobios Publication, Delhi, 2002.

H. Panda: *Handbook on Milk and Milk Proteins*, Asia Pacific Business Press, Delhi, 2011.

H.A. Modi: *Fermented Milk Products*, Aavishkar Publication, Delhi, 2011.

J.G. Davis: *Milk Testing : The Laboratory Control of Milk*, Agrobios Publication, Delhi, 2010.

Mangala Kango: *Milk and Milk Products*, RBSA Publication, Delhi, 2006.

Manish L Srivastava: *Handbook of Milk Microbiology*, Daya Publication, Delhi, 2002.

Mohammad Rasheed; *Milk Hygiene and Safety*, Jaya Publication House, Delhi, 2015.

Mohammad Raziuddin and Ashok Hembade: *Milk and Milk Products Technology*, Jaya Publication House, Delhi, 2013.

Mohd. Kalimuddin: *Milk Analysis and Dairy Farming*, Enkay Publication, Delhi, 2012.

Mohd. Kalimuddin: *Milk and Milk Products*, Enkay Publication, Delhi, 2012.

N N Godbole: *Milk : The Most Perfect Food*, Biotech Books Publication, Delhi, 2007.

Rajendra Kumar Pandey: *Indigenous Milk Products : Vol 1: Fat Rich Milk Sweets*, Biotech Books Publication, Delhi, 2014.

S. Rajagopal and S.K. Roy: *Milk and Milk Products Technology*, BS Publications, Delhi, 2014.

S.N. Mahindru: *Milk and Milk Products*, APH Publication, Delhi, 2012.

Shivashraya Singh: *Milk and Milk Processing: Vol.1: Dairy Technology* , New India Publication Agency, Delhi, 2014 .

Varnam: *Milk and Milk Products: Technology, Chemistry, and Microbiology*, Springer Publication, Delhi,2010.

William Barbour Nevens: *Principles of Milk Production*, Axis Books Publication, Delhi, 2010.

Index

L

M

O

P

R

S

T

U

V